RWTHedition

Series editor

RWTH Aachen University, Aachen, Germany

More information about this series at http://www.springer.com/series/7858

Reinhart Poprawe · Konstantin Boucke
Dieter Hoffman

Tailored Light 1

High Power Lasers for Production

Springer

Reinhart Poprawe
Fraunhofer-Institut für
Lasertechnik (ILT)
Aachen
Germany

Dieter Hoffman
Fraunhofer Institute for Laser
Technology (ILT)
Aachen
Germany

Konstantin Boucke
Fraunhofer-Institut für
Lasertechnik (ILT)
Aachen
Germany

ISSN 1865-0899 ISSN 1865-0902 (electronic)
RWTHedition
ISBN 978-3-642-01233-4 ISBN 978-3-642-01234-1 (eBook)
https://doi.org/10.1007/978-3-642-01234-1

Library of Congress Control Number: 2016956468

Printed on acid-free paper

This Springer imprint is published by the registered company Springer-Verlag GmbH, DE part of Springer Nature
The registered company address is: Heidelberger Platz 3, 14197 Berlin, Germany

Preface

This book is the first of a twinset, displaying the fundamental characteristics of lasers and laser applications. Whereas "Tailored Light 2" focusses on the applications, this volume is dedicated to the laser sources in use and their potential in high-power applications.

The sources of coherent photons, lasers, are fascinating because of their unique borderline characteristics. Borderline in the sense of fundamental limits of our natural science laws, as they operate at the so far ultimately highest speed ever achievable in this whole universe, the speed of light. We know that any form of matter would need an infinite amount of energy to be accelerated to such speed limits, however, photons have no property of mass and thus are able to move at the speed of light—again: with no mass, no inertia, no limits of the materialized world we live in. Are there speeds beyond that limit? We do not know, but we do know the borderline, this cornerstone of wonderful universally unique properties and this highest imaginable quality of energy, massless, physically unlimited in density and ultimately fast.

Who Should Read in this Book?

The book should be read by individuals involved in innovative processes based on technology in general, because of its applied context. Also the students of photonics or laser technology will find valuable context rather on the fundamental end of science.

Technological innovation is increasingly characterized by high complexity of content and the related processes need systematic structuring. People active in innovation processes and have developed an interest in laser technology do not need to know about rate equations or the details of "Light Amplification of Stimulated Emission of Radiation" (LASER), but they should know about the vast variety of lasers and the application potential, the "what", i.e., there are different wavelengths suited for different processes and different materials, e.g., glass transmits 1.3 μm perfectly; hence lasers with such wavelengths will be used for information, communication and internet around the globe. For glass cutting highly absorbed far-infrared wavelengths are used, but also other wavelengths are used for which glass is extremely transparent at low intensities, however, are also absorbed extremely, if utilized at critical intensities.

You do not need to know how and which laser medium is applied for generating which pulse lengths, but you should know, there are fs lasers (10^{-15} s) fantastically suited for ultra-precision machining available today up to kW of average power, thus being relevant for modern manufacturing. And you do not need to know in digital photonic production how to "slice" a 3-D-design from a computer into process data for "3-D-printing," but the relevance of high-power diode lasers and their potential for power and cost scaling by automated production enabling what among others "The Economist" calls the third industrial revolution.

Student, engineer, academically or industrially active scientist, and advanced technologist however will need and want to know "why that," "why now," and "how" and thus will benefit from the context of this book. The details and fundamentals of the different lasers in terms of active media (i.e., gases, solid-state materials, and semiconductors) as well as excitation processes, resonator designs, and system characteristics are of great relevance. There will be questions arising on how to design certain properties of laser radiation and why certain concepts of material selection, design of geometry, and resonators will be most suited. These questions will be answered in this book.

As a bridge between the world of science (bottom up) and the impact of laser technology on our societal challenges like mobility, information technology, health, energy, or security (top down) "Tailored Light" connects markets and technologies, core competencies, and business opportunities. The systematics of transdisciplinary innovation cannot be addressed explicitly in detail and would be beyond the scope of consideration here, but for the example of laser technology the links and cross-fertilizing opportunities of societal ad thus economic and ecologic relevance can be deducted.

Aachen, Germany Reinhart Poprawe
April 2016

Contents

Chapter 1
The History of Laser

The history of laser begins in the year 1960, when THEODORE MAIMAN reported about pulsed laser activity of a ruby laser for the first time. Up until this important point of time, numerous cutting-edge discoveries and knowledge were necessary. For instance, Maiman's realization of the first laser drew upon physical models made by SCHAWLOW and TOWNES on the amplification of radiation in the optical wavelength range (1958) and upon other work.

Since a laser functions in a manner based on amplifying electromagnetic radiation in an appropriate medium, understanding the interaction between radiation and matter was decisive for the discovery of the laser principle. A prerequisite for this understanding was, on the one hand, an appropriate description of radiation and, on the other, a detailed model of matter.

For most areas, an appropriate description of electromagnetic radiation was formulated by MAXWELL in 1873: the Maxwell theory of electromagnetic radiation, based on *Maxwell's equations*. This point of time may seem to be chosen rather arbitrarily as representing the beginning of the prehistory of laser (Table 1.1). A reliable model of matter was made possible by the development of *quantum mechanics* at the beginning of the twentieth century. The quantum mechanical model of the atom revolutionized physics. Together with this model, the first detailed representations of the interaction between radiation and matter originated.

Yet it was not until 1951 when the basic idea of the laser, the idea of amplifying electromagnetic radiation by matter, matured. From here on, up until the theory of the maser and the laser, only a very small, last step was missing: the idea of self-amplification by feedback coupling. MEISSNER had already formulated the *principle of feedback coupling* by 1913. The correct combination of these three basic building blocks—the theory of electromagnetic radiation, quantum physics' model of the atom, and the principle of feedback coupling—lead to the discovery of the laser.

R. Poprawe et al., *Tailored Light 1*, RWTHedition,
DOI 10.1007/978-3-642-01234-1_1

Table 1.1 The prehistory of laser

1873	Electromagnetic field and light theory (Maxwell)
1887	Photoelectric effect (Hertz)
1888	Proof of electromagnetic waves (Hertz)
1890	Series laws of the atomic spectra (Rydberg)
1893	Open resonator (Thomson)
1896	Discovery of X-ray radiation (Röntgen)
1900	Quantum hypothesis and radiation law (Planck)
1901	Proof of radiation pressure (Lebedev; Nichols and Hull)
1902	Proof of wavelength dependence of photoemission (Lenard)
1905	Light quantum hypothesis (Einstein)
1909	Statistics of radiation and wave–particle duality (Einstein)
1911	Statistics of identical quanta (Natanson)
1912	Proof of wave character of X-ray radiation (v. Laue; Friedrich and Knipping)
1913	Discovery of feedback principle (Meißner) Bohr's atomic model and correspondence principle (Bohr)
1914	Excitation of a spectral line through electrons of discrete energy (Franck and Hertz)
1917	Postulate of induced emission (Einstein)
1922	Proof of quantization of angular momentum with silver atoms (Stern and Gerlach) Doppler frequency shift during scattering of X-ray beams upon electrons (Compton)
1923	Confirmation of light quantum hypothesis by explanation of Compton effect as collision problem (Compton; Debye) Postulate of wave properties of material (de Broglie)
1924	Bose–Einstein statistics of identical quanta (Bose; Einstein)
1925	Quantum mechanics (Heisenberg), Exclusion principle (Pauli) Hypothesis of electron spin (Goudsmit and Uhlenbeck)
1926	Representation of quantum mechanics by matrices or more generally by noncommutative algebra (Born, Heisenberg and Jordan; Dirac) Fermi–Dirac statistics of identical quanta (Fermi; Dirac) Wave theory of material and its equivalence with quantum mechanics (Schrödinger)
1927	Principle of uncertainty (Heisenberg) Proof of wave properties of electrons by interference during the reflection upon crystals (Davisson and Germer; Thomson)
1928	Relativistic wave equation of the electron (Dirac) Proof of induced emission (Ladenburg and Kopfermann)
1929	Quantum field theory (Heisenberg and Pauli)
1930	Theory of spontaneous emission from lamps (Weisskopf and Wigner)
1933	Proof of radiation momentum (Frisch)
1934	Quantum theory of scalar fields (Pauli and Weisskopf)
1936	Proof of radiation angular momentum (Beth)
1948	Proposal for holography (Gabor)
1950	Proof of inversion with the nuclear spin system of LiF (Purcell and Pound)
1951	Proposals to amplify electromagnetic radiation by population inversion of a medium and stimulated emission (Fabrikant; Townes)

(continued)

Table 1.1 (continued)

1953	Proposal for maser (Weber)
1954	Maser principle (Basov, Prokhorov) First maser with NH_3 molecules (Gordon, Zeiger, and Townes)
1955	Three level scheme in gases (Basov, Prokhorov)
1956	Three level scheme in solid bodies (Bloembergen)
1957	Solid-state maser (Feher, Bordon et al.)
1958	Ruby maser (Makhov et al.) Patent for semiconductor maser (Heywang) Proposals to amplify in the optical range (Schawlow and Townes; Prokhorov; Dicke)
1959	Patents on laser apparatus with population inversion by optical pumping or electrical discharges in the laser medium (Gould) Proposal for gas laser (Javan) Proposal for semiconductor laser (Basov, Bul, and Popov)

After 1960 the development of the laser itself occurred in rapid steps (Table 1.2). Nonetheless, nearly three decades passed before the laser could gain a foothold as an industrial tool. While this time span may seem long, a comparison to the development of other technologies—as computer technology, for example—shows that 25 to 30 years is quite typical for them to mature. Within these three decades, three development stages are traversed. At first, foundations and an understanding for the physics of new technologies are developed. Based on this, laboratory

Table 1.2 The history of laser

1960	First pulsed laser activity: ruby solid-state laser (Maiman) Rate equations (Statz and de Mars) Proposal of excimers as a laser medium (Houtermans) First observation of relaxation oscillations (Collins, Nelson; Schawlow et al.)
1961	First continuous laser activity: HeNe laser at 1.15 μm (Javan, Bennett, and Herriott) Resonator theory (Fox and Li) Proposal of a chemical laser (Polanyi) Nd:Glass laser (Snitzer) Proposal of a laser with light guide cavity (Snitzer) Frequency doubling by nonlinear effects (Franken, Hill, Peters, and Weinreich) First proposal for a Q-switched laser (Hellwarth)
1961/62	Realization of holography by means of a laser (Leith and Upatnicks)
1962	First semiconductor laser (Hall et al.; Nathan et al.; Quist et al.) Q-switched ruby laser (McClung and Hellwarth; Collins and Kisluik) Confocal resonator (Boyd and Kogelnik) Stimulated Raman emission (Eckhardt et al.) Phase matching for frequency generation (Giordmaine) Optical demodulation (Bass, Franken, Ward, and Weinreich) HeNe laser at 632.8 nm (White and Rigden) Parametric amplification (Kingston; Kroll)

(continued)

Table 1.2 (continued)

1963	Photon statistics and quantum theory of coherence (Glauber; Sundarsham) Semiclassical laser theory: coexistence and competition of modes (Haken and Sauermann; Tang et al.; Lamb jr.) Proposal of the gas dynamics laser (Basov and Oraevskij) Fiber laser (Elias Snitzer)
1964	Frequency tripling (Maker, Terhune, and Savage) Proposal and calculation of waveguide gas lasers (Marcatili and Schmeltzer) Quantum theory of the lasers and nonlinear theory of the coherence properties of laser light (Haken) Argon- and krypton-ion laser in the visible range (Bridges) HCN laser with 377 μm in the gap between infrared and microwaves (Gebbie, Stone, and Findlay) Continuous CO_2 laser at 10.5 μm and 9.5 μm (Patel) Nobel Prize for Physics to Townes, Basov, and Prokhorov Stimulated Brillouin scattering (Bret et al.; Benedek et al.; Brewer) Nd:YAG laser at 1064 nm (Geusic, Marcos, and van Vitert) Laser activity by excited iodine atoms produced by photochemical dissociation (Kasper and Pimentel)
	Q-switching by saturable absorbers (Goodwin et al.; Kafalas et al.; Miller; Sorokin et al.) Proposal of compression by chirp pulses (Gires and Tournais) Prediction of self-focussing (Chiao, Garmire, and Townes)
1965	Prediction of photon statistics at the laser threshold (Risken) First chemical laser with HCl at 3.8 μm (Kasper and Pimentel) Color center laser with KCl:Li/F_A (Fritz and Menke) Tunable parametric light oscillator (Giordmaine and Miller) Density matrix equations for lasers (Weidlich and Haake) Observation of self-focussing (Pilipetskii and Rustamov; Lallemand and Bloembergen; Hauchecorne and Mayer) Mode locking to produced picosecond pulses in solid-state lasers (Maker and Collins)
1966	10 kW gas-dynamic CO_2 laser (Kantrowitz et al.) Pulsed dye laser (Sorokin and Lankard, Schäfer, Schmidt and Volze)
1968	Proposal of compression of chirp pulses via group velocity dispersion or dispersive delay chains (Giordmaine, Duguay, and Hansen) Picosecond pulses by mode locking of a dye laser (Schmidt, Schäfer)
1969	Prediction of optical bistability (Szöke, Danev et al.) Pulse compression gratings (Treacy) Industrial installation of three lasers for automobile application (G M Delco)
1970	Self-phase modulation in crystals, fluids, and glasses (Alfano and Shapiro) Pulsed TEA-CO_2 laser (Beaulieu) Continuous dye laser (Peterson, Tuccio, and Snavely) Excimer laser at Lebedev Labs Moscow (Nikolai Basov, Yu M. Popov) CW semiconductor laser at Ioffe Physico-Technical Inst. & Bell Labs, (Alferov' Group, Mort Panish, Izuo Hayashi)
1971	Distributed feedback dye laser (Kogelnik and Shank) Proposal of a free-electron laser (Madey) Excimer-Xe_2 laser (Basov, Danilychev, and Popov)

(continued)

Table 1.2 (continued)

1972	Nonlinear optical phase conjugation (Zel'dovich et al.; Nosach et al.)
1974	Prediction of optical transistor properties (McCall)
1975	Noble gas halide-excimer laser (Searles and Hart; Ewing and Brau) Compression of mode lock dye laser pulses through lattice pairs (Ippen and Shank) Deterministic chaos in the laser (Haken)
1976	Proof of optical bistability and transistor function (Gibbs, McCall, and Venkatesan)
1977	Realization of the free-electron laser (Deacon, Elias, Madey et al.)
1981	GVD and self-phase modulation in fibers, pulse compression
1982	30 fs pulses (Shank, Fork, Yen, Stolen, and Tomlinson) Titanium Sapphire laser at MIT Lincoln Labs (Moulton)
1987	6 fs pulses (Fork, Brito Cruz, Becker, and Shank)
	Erbium fiber amplifier (Payne)
1991	ThinDisk laser, proposal of the concept (Giesen, Wittig, Brauch, Voß)
1994	InnoSlab laser, proposal of the concept (Du, Loosen)
1996	Petawatt laser demonstration at Lawrence Livermore National Labs.
1997	1 kW cw diode-pumped rod laser at Mitsubishi (Takada et al.) and ILT Aachen (Poprawe, Hoffmann et al.) Atom laser at MIT Lincoln Labs (Ketterle)
2000	1 kW cw ThinDisk laser (Giesen et al.)
2001	Isolated attosecond pulses (Hentschel, Kienberger, Krausz et al.) 2 kW fiber-coupled diode laser by Jenoptik (Dorsch, Hennig et al.)
2003	1 kW cw fiber laser by SPI (Y. Jeong, J.K. Sahu, D. N. Payne, and J. Nilsson)
2004	1 kW cw InnoSlab (Schnitzler et al.)
2006	Silicon laser (Bowers)
2009	10 kW single-mode cw fiber laser by IPG (Gapontsev et al.)
2010	10 Petawatt laser by Lawrence Livermore National Labs (M. Perry et al.) 1 kW average power fs laser at ILT (Rußbüldt et al.)
2011	10 kW fiber-coupled diode laser by Laserline (Krause et al.)
2012	10 kW fiber-coupled single ThinDisk laser by Trumpf (Killi et al.)

systems are constructed and refined technically. Using these systems, industrial applications of the new technologies are, in turn, fathomed.

Laser technology has largely passed through these development stages. In many areas of fundamental research, the laser has become an indispensable instrument. Interference measurements for ultraprecise length or velocity measurement or the ultra-short-time spectroscopy with temporal resolution in the femto-second scale exemplify the laser's significance in research today.

The numerous applications in research have prepared the laser's entrance into the industry. First applications in production technology were the drilling of hard working materials such as diamond and sapphire, the joining of microelectronic components, and the cutting of steel sheets. In the meantime, a large amount of industrial laser applications have been developed.

Laser technology is currently maturing into a standard technology. Still at the beginning of its wide establishment in the industry, laser technology has a huge potential for further development.

Footnote: Readers are encouraged to communicate additional information on further fundamental or industrial "first times."

Chapter 2
An Introduction to Laser Technology

Before addressing the foundations and the functions of the laser in detail in the following chapters, we would like to give an overview of its most important concepts and interrelations. In this way, the arrangement of the details discussed later shall be simplified within the overall picture.

2.1 The Laser

Originally the term "laser" was the abbreviation for *Light Amplification by Stimulated Emission of Radiation*. Therefore, this term initially characterized a process, namely, a special kind of light amplification. Subsequently, the term "laser" was increasingly used to characterize the technical device that makes use of this process. With this in mind, "laser" thus characterizes a special source of radiation (Fig. 2.1).

Laser radiation exhibits special properties, which render it interesting for various technical and scientific applications. The cause for its properties rests in the way it is produced, or rather amplified through the *laser process*: in the physical process of *stimulated emission*. A special construction of the laser source is necessary, however, to be able to use the stimulated emission and to actually obtain laser radiation.

In the meantime, a wide variety of different laser types have been developed. They vary greatly in their dimensions, output power, and emission frequency. The smallest lasers—the *semiconductor lasers*—have dimensions in the submillimeter range and a typical output power of several milliwatts. *High power lasers* for industrial applications produce continuous output power of up to 40 kW and have dimensions ranging several meters. The lasers used in fusion research are up to

R. Poprawe et al., *Tailored Light 1*, RWTHedition,
DOI 10.1007/978-3-642-01234-1_2

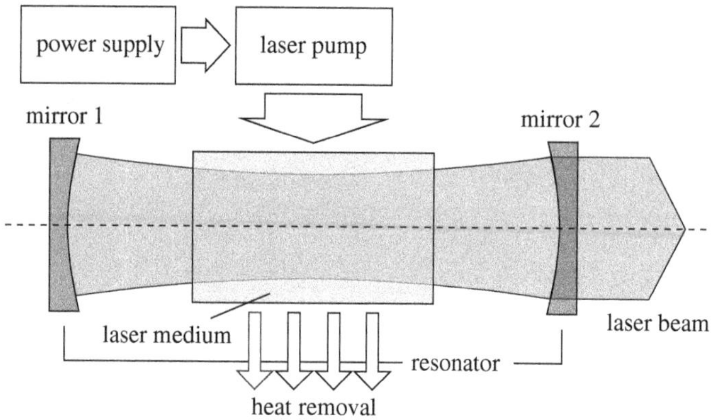

Fig. 2.1 Schematic representation of the essential components of a laser

100 m long and provide short laser pulses with extremely high energy. With these lasers the output power can reach 1 TW[1] for a brief period of time.

2.1.1 Stimulated Emission

Common to all of the lasers is the *laser principle*, which is based on the *stimulated emission* of radiation. The process of *stimulated emission* was postulated by EINSTEIN in 1917, as a complement to the already known processes of *spontaneous emission* and *absorption*. These three elementary processes describe the interaction between radiation and material.

Material consists of atoms. Without going into atomic structure in detail, it can be assumed that every atom can be found in various *states*, which can be distinguished from each other by the inner energy of the atom. The concept of the atom from quantum mechanics leads, moreover, to the proposition that only states with discrete energy are possible. These states are also designated *energy levels*, the level with the lowest possible energy being called the *ground state*.

To begin with, light is characterized by its frequency and intensity. Light also has a specific energy content, and as in the case of the atom, quantum theory says that only discrete energies are possible for a light field with a given frequency. The energies possible always differ by the same, elementary amount of energy. These elementary amounts of energy are also designated as a *light quantum* or *photon*. The energy of a light quantum is proportional to the frequency ν of light:

$$E_{\text{Photon}} = h\nu. \tag{2.1}$$

[1] $1\ \text{TW} = 10^{12}\ \text{W}$.

h Planck's constant

On the other hand, the intensity I of light is defined by the number of light quanta,

$$I \sim n_{\text{Photon}}. \tag{2.2}$$

n_{Photon} Number of light quanta or photons.

With these models of atoms and light, the interaction processes can be understood. What is easiest to grasp phenomenologically is the process of absorption. When light hits a material obstruction, the light's intensity is reduced, or rather, the number of photons is reduced: the photons are absorbed by the atoms. In this process, the energy of the absorbed photons is also transferred to the atom, so that the atom subsequently is found in a higher energy level (Fig. 2.2). This is designated as the *excitation* of the atom. Since the atom only has specific, discrete energy levels, the energy of the photon has to correspond—this is the presupposition for the process—to the energy gap to the atom's next energy *level:*

$$h\nu = E_m - E_n. \tag{2.3}$$

$E_{m,n}$ Energy levels of the atom.

Emission is the reversal process of absorption: an excited atom moves from a higher energy level to a lower one and emits the released energy in the form of a photon.

There are two different emission processes. In the case of *spontaneous emission*, the process described occurs without the atom being influenced from outside. This is the emission process prevailing in nature. A hot body, for example, cools down by releasing heat radiation, whereas its atoms return from excited levels to the ground state.

For the *stimulated emission*, however, the emission process is triggered by the impact of a light quantum fitting in frequency. The light quantum that impacts is

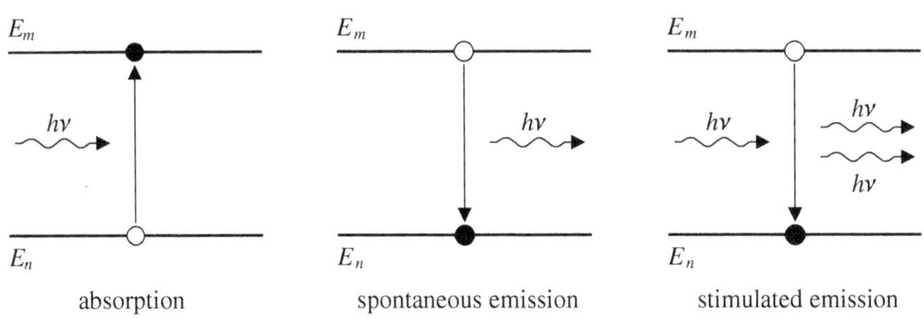

Fig. 2.2 Schematic representation of the elementary interaction *processes* of an atom with light

thus not absorbed in this case, but rather causes the emission of a further light quantum. Decisive for the functionality of a laser is that the second emitted light quantum corresponds to the exciting light quantum in its frequency, phase, and emission direction. This means that the incident light wave is amplified by this process, since an additional, identical light quantum is added.

2.1.2 Population Inversion and Amplification

In a medium consisting of very many atoms, all the three interaction processes occur simultaneously. Stimulated emission has to become the dominant process to attain an overall amplification of the light source.

In thermal equilibrium, most of the atoms are found in the ground state, since the spontaneous emission—in absence of a light source—ensures that the atoms from excited, higher energy levels return to lower ones.[2] The Boltzmann equation is valid for the population numbers of the different energy levels (Fig. 2.3):

$$E_n > E_0 \colon N_n = N_0 \exp\left(-\frac{E_n - E_0}{k_B T}\right), \quad \sum_{n=0}^{\infty} N_n = N_{Total}. \tag{2.4}$$

$E_{n,}$ N_n	Energy and population number of the nth energy level.
$k_B = 1381 \times 10^{-23}$ J/K	Boltzmann's constant.
T	Absolute temperature.
N_{Total}	Total number of observed atoms.

The transition rate of each elementary process is proportional to the population number of the initial energy level, since the population of this level is a prerequisite for this process to occur. This means that the absorption process constantly outweighs the stimulated emission in thermal equilibrium, since more atoms are located in the energetically lower level. This corresponds to daily experience, according to which light is weakened when it penetrates a body.

So that the stimulated emission exceeds the absorption in the *laser medium*, more atoms have to be found in the upper laser level[3] than in the lower. This so-called *population inversion* has to be produced (Fig. 2.4). If it has been reached, then the predominance of stimulated emission over absorption leads to an amplification of the light wave, which is proportional to the population inversion.

[2]In addition to spontaneous emission, there are a number of further *relaxation processes*, which, however, will not be addressed here.

[3]Laser levels are those energy levels between which the desired transition occurs for the laser emission.

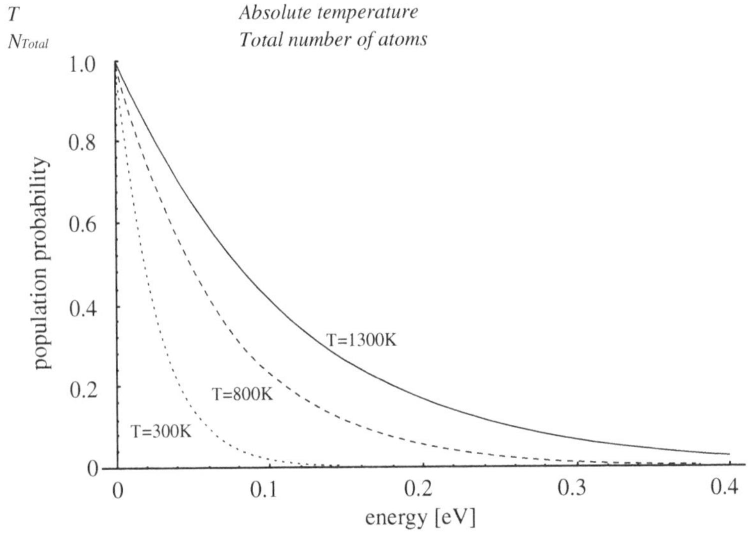

Fig. 2.3 The Boltzmann distribution for three different temperatures. The energy is given in electron volts: 1 eV = 1.6×10^{-19} J

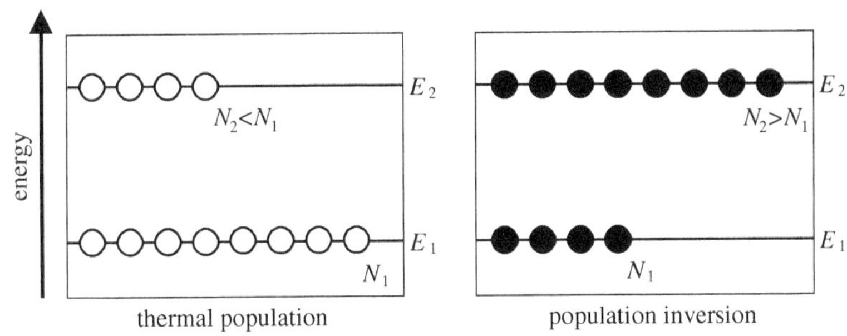

Fig. 2.4 For thermal population the population number constantly declines with increasing energy. With inversion, however, the population number is larger at a level of higher energy than at a lower level

The population inversion has to be generated by an external excitation process, called the *pump mechanism*.

The spontaneous emission presents a *competing process* for the stimulated emission: the light quanta released by spontaneous emission do not contribute to the amplification process, since they are emitted in the medium isotropically[4] in every direction and with any phase. Therefore, spontaneous emission represents a *loss channel*, since it degrades the population inversion without producing amplification.

[4]Isotropic: distributed equally.

The probability of transition for spontaneous emission is not dependent on the amount of light quanta present. This probability only reflects the *lifetime* of the atom in the upper laser level[5]; the longer the atom can remain in the upper laser level in the medium, the lower the transition probability apparently is. On the contrary, the probability for stimulated emission increases proportionally to the number of light quanta, since the presence of a corresponding light quantum initially leads to a stimulation of this process. Above a specific *threshold intensity*, the stimulated emission thus outweighs spontaneous emission, and the prerequisites for amplification are given.

As the intensity increases along the propagation axis of the light in the laser medium, it follows an exponential growth formula. The increase in intensity is proportional to the transition rate of the stimulated emission, which in turn is proportional to the intensity itself:

$$\frac{\mathrm{d}}{\mathrm{d}z} I = (g - \alpha) \cdot I \quad \Rightarrow \quad I(z) = I(0) \cdot \mathrm{e}^{(g-\alpha)z}. \tag{2.5}$$

I(z) Light intensity after propagating over the distance z in the laser medium
G Amplification coefficient
α Absorption coefficient

Since losses constantly arise due to absorption and spontaneous emission, the proportionality constant is made up of the *amplification* and the *absorption coefficients*.

2.2 The Laser Medium

The substance used for the optical amplification is designated as the *laser medium* (Fig. 2.5). For this, substances in all the aggregate states can be used: solid bodies, liquids, gases, and plasma. As a rule, the laser medium used names the laser: for the CO_2 laser, carbon dioxide (CO_2) is used as the laser medium, in the ruby laser, a ruby crystal is used, and in the helium–neon laser, a mixture of the gases helium and neon.

The decisive criterion for a laser medium is that a maximum amount of population inversion is attained in a simple manner. For this two prerequisites have to be fulfilled:

- An energy level appropriate as the upper laser level, with a preferably long lifetime, and
- An appropriate and sufficiently efficient pumping mechanism.

[5]What is meant by *lifetime* of an energy level is the time that the uninfluenced atom remains at this level before it passes into a lower energy state.

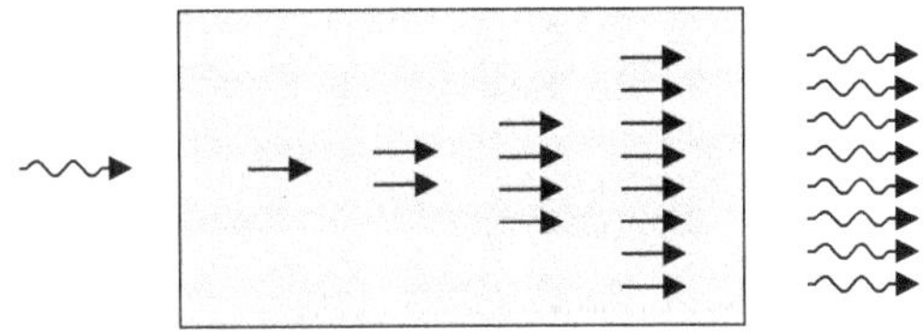

Fig. 2.5 The laser medium amplifies the radiation passing through it

For continuous laser operation, the pumping mechanism has to maintain the population inversion during laser operation. For this, at least the inversion degradation by both emission processes has to be counterbalanced. The larger the lifetime of the upper laser level is, the lower the emission rate is, hence making a lower pump power sufficient to guarantee laser operation.

Through the structure of its energy levels, the laser medium defines the useable laser frequencies. As a rule, amplification with a medium can only be attained with few, discrete frequencies. A large number of different laser media, therefore, are necessary to provide laser radiation of appropriate frequencies for all of the numerous applications.

The output power and the temporal behavior of a laser are determined by the laser medium and the pumping process. Since the latter provides the energy to be transformed into laser radiation, the mean laser power cannot be larger than the pump power. For several gas and solid-state lasers, this mean power typically amounts to only a few percent of the pump power; with semiconductor lasers, it rises to more than 60 %. The pump power, however, does not have to be transformed into laser radiation continuously. In *pulsed lasers*, the pump power is accumulated and released in individual, but significantly more intensive laser pulses.

2.2.1 The Laser Pumping Process

The laser pumping process creates a population inversion in the laser medium (Fig. 2.6). For this, the laser medium has to be excited, which means energy has to be supplied to it. To do this, the following mechanisms are used:

- Gas discharges for gaseous laser media,
- Optical excitation through flash lamps or a *pump laser* for solid-state lasers, or
- Pumping by electrical current for semiconductor lasers.

Through pumping, atoms are excited from the lower into the upper laser level. As a rule, the excitation does not, however, occur via a direct transition between the laser levels, but rather indirectly via additional energy levels.

Figure 2.7 presents a schematic diagram of a *four-level system*. Through the pumping process, a higher level is filled, beginning from the ground level. In an ideal case the lifetime in this higher level is very short, and the atoms transfer quickly into the level below, the upper laser level. From there the laser transition

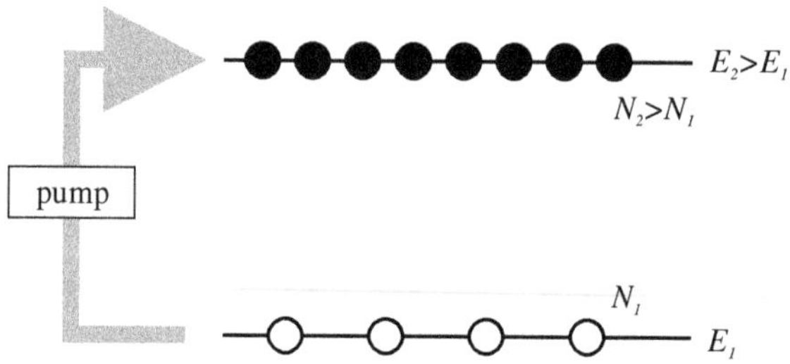

Fig. 2.6 The pumping mechanism serves to produce a population inversion. For this, atoms from a lower energy level are excited to a higher one

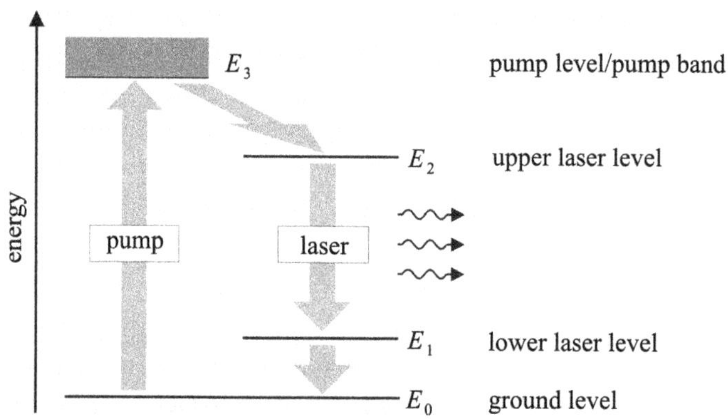

Fig. 2.7 Diagram of a four-level system. Short lifetimes in the states E_1 and E_3 result in the population inversion between the laser levels being easily attained, on the one hand, and the pumping process working from a full into an empty level

occurs in the lower laser level, which also exhibits a very short lifetime, so that the atoms quickly return to their ground state.

On the one hand, this results in the population inversion between the laser levels E_1 and E_2 being easily attained, since the upper laser level is quickly filled and the lower quickly emptied. On the other hand, the pumping process can also work very effectively, since the ground level is quickly refilled, while the upper pump level E_3 remains almost empty on account of its short lifetime.

In general, the four-level system cannot, however, be realized within an individual atom. As a rule the pumping process and the laser transition take place in different atoms,[6] whereas the energy transfer from one atom to others occurs via collisions of the atoms themselves.

[6]For example, in a helium–neon (HeNe) laser: pumping process using helium atoms, the laser transition using neon atoms.

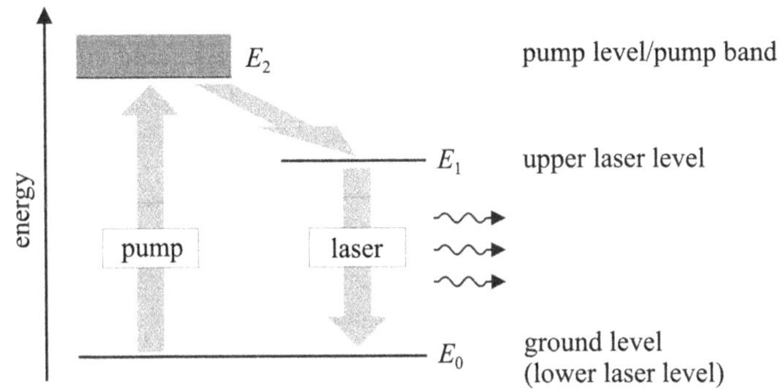

Fig. 2.8 Diagram of a three-level system. Ground level and lower laser level are identical

Several lasers work using a *three-level system*. For this, the lower laser level is identical to the ground state (Fig. 2.8). The disadvantage of the three-level system is the higher pump power generally needed to reach inversion; in exchange, the three-level system can be realized within a single atom.

The details of the pumping process and the transition processes between the energy levels depend greatly upon the laser medium used. The energy levels can be atomic or molecular, or also energy bands of a solid body. The transitions between the levels can be triggered in very different ways, for example, by

- Light irradiation,
- Collisions between atoms,
- Excitation of vibrations of molecules or solid bodies, and
- Chemical reactions.

Chapter 10 will address the details, which are specific to each laser family.

2.2.2 *Cooling*

In order to reach the maximum amplification within the laser medium, population inversion has to be as large as possible. It can be enlarged, on the one hand, by increasing the pump power. This results in a higher population of the upper laser level. If the lower laser level is not identical with the ground level, a second possibility consists in reducing the population of the lower laser level by lowering the temperature. Thereby, the population probability in thermal equilibrium falls in the higher energy levels (cf. Fig. 2.3), and a correspondingly larger inversion is hence enabled.

The cooling of the laser is thus very significant for the amplification process itself. Accordingly, when a laser system is being developed, much emphasis is placed on effectively dissipating the heat arising in the laser medium: In some gas laser systems, the laser gas is pumped through the resonator at nearly the speed of sound, and before being directed back, is cooled down in high-performance cooling units. For semiconductor lasers, the heat is transported away using microstructured, water-cooled heat sinks.

2.3 Feedback and Self-Excitation

Up to this point, along with the laser medium and pumping process, an optical amplifier was introduced, which amplifies the input signal through stimulated emission (**Fehler! Verweisquelle konnte nicht gefunden werden.** Fig. 2.9). To reach the *light source* from the *light amplifier*, the system has to be independent of an input signal. In addition, the laser should take advantage of the maximum amplification of the laser medium. To attain both of these points, *the feedback principle* is used.

The feedback principle is a general physical principle, which was discovered by A. Meissner in 1913. It serves to stabilize the amplification of oscillations of a specific frequency. Any input signal is coupled into an amplifier. The *feedback* consists in a part of the amplified signal being conducted again to the amplifier input and being amplified anew (Fig. 2.10). The rest is outcoupled and is available as output signal (Fig. 2.9).

The amplifier with feedback represents a *self-oscillating system* (oscillator), which is excited by the input signal. Through variations of delay of the feedback, the

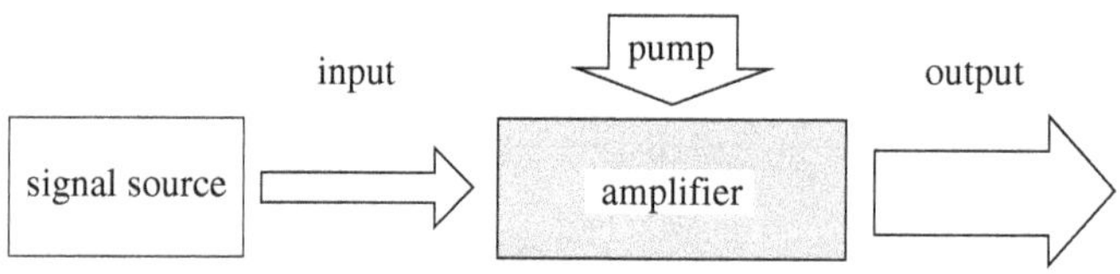

Fig. 2.9 The simplified amplification process

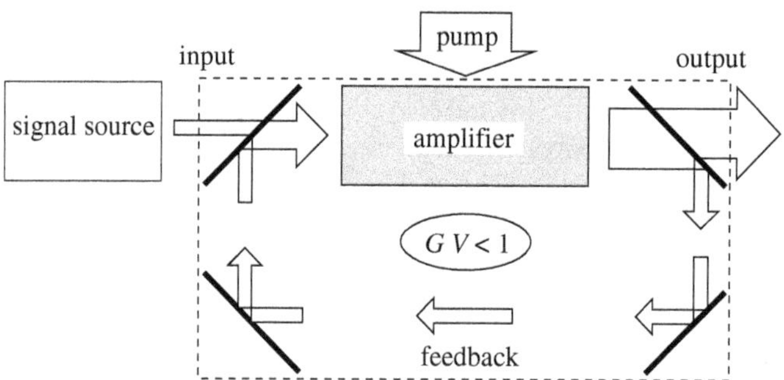

Fig. 2.10 The amplification process with feedback. A part of the amplified signal is fed back into the input. The system functions as an amplifier

system can be tuned to a specific oscillation frequency. The temporal delay provides the phase of the feedback signal at the input. With feedback of *equal phase* (*positive feedback*), the oscillator is in *resonance*. This is the case when the *phase condition*

$$\Delta T_{feedback} = n \cdot \frac{2\pi}{\omega} \Rightarrow \omega = n \cdot \frac{2\pi}{\Delta T_{feedback}}, \quad n = 0, 1, 2, \ldots \tag{2.6}$$

$\Delta T_{\mathrm{feedback}}$ Delay of the feedback.
ω Resonance frequency of the oscillator.

is fulfilled. An input signal of this frequency experiences an additional amplification through the feedback, whereas the nonresonant frequencies are amplified less because of the phase difference to the feedback signal. For a phase difference of π, the feedback signal counteracts the input signal exactly, such that even weakening results instead of amplification. This case is called an *inversely phased* coupling (*inverse feedback*). In this way, a *stabilization* of the amplification is attained against the influence of noise from the amplifier.

As long as the feedback share of the signal is small, the amplifier with feedback represents a *damped* oscillator. If the input signal is turned off, the signal power in the system decays, since the amplification of the feedback alone is not enough to balance the outcoupling losses.[7] If the feedback share of the signal is continuously increased, however, the *threshold to self-excitation* will be reached: at this point the losses by the outcoupling are completely compensated for by the amplification of the feedback signal. This is described by the equation

$$G \cdot V = 1. \tag{2.7}$$

G Amplification factor (gain).
V Loss factor in the feedback cycle.

The consequence is that the power circulating in the system remains constant after the input signal is switched off, and the output signal becomes independent of the input signal. This process is called *self-excitation*. The feedback amplifier with self-excitation oscillates free of damping.

Above the threshold for self-excitation, the amplitudes of resonant signal frequencies increase until the saturation limit of the amplifier is reached. The amplification is, thus, not linear in this area: small signal amplitudes experience a greater amplification than large signal amplitudes, such that an *amplitude-stabilized output signal* results.

The laser is a self-excited oscillator. So that laser operation can begin, the threshold for self-excitation (Eq. 2.7) has to be reached; in the case of a laser, this threshold is also called the *laser threshold*. What is dispensed with is the incoupling

[7]With regard to the feedback cycle, the outcoupling represents a loss since it reduces the power circulating in the system.

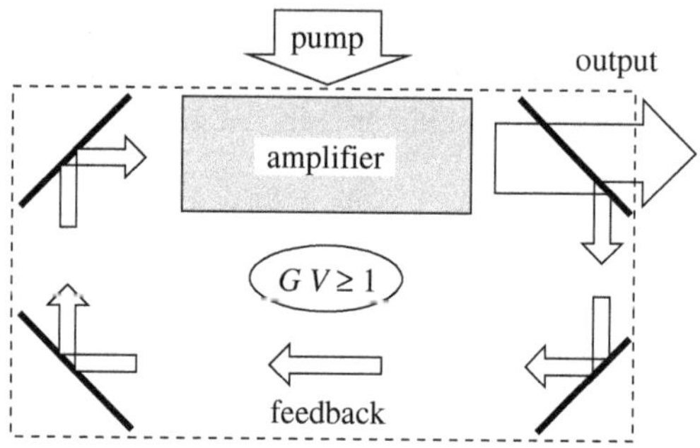

Fig. 2.11 Feedback and self-excitation as in the case of a laser. The input signal is dispensed with, so that now a signal source exists instead of a signal amplifier

of an input signal to trigger the self-excitation (Fig. 2.11). Amplification and feedback are set in motion by a disturbance in the amplifying laser medium, the spontaneous emission. This incidental signal is fed back and amplified by stimulated emission. When the laser threshold is exceeded, the stimulated emission becomes the predominant process—one says here that the laser begins to oscillate out of the noise (Fig. 2.12).

2.4 The Laser Resonator

The *laser resonator* is the laser component responsible for the feedback. In the simplest case, it consists of two mirrors. In the inside of the resonator, between the mirrors, the laser medium is found (Fig. 2.13). At least one of both mirrors is partially transparent so that a share of the amplified radiation is outcoupled as *laser radiation*.

The phase condition of the feedback results in only light waves of specific frequencies, the so-called *eigenmodes* or *resonator modes,*[8] developing in the resonator. The *eigen* modes represent the resonant modes of oscillation of the electromagnetic field in a resonator. *Thus the resonator determines the possible laser frequencies*. The laser medium amplifies only eigenfrequencies near the frequency of the laser transition. *The final laser frequency is the eigenfrequency of the resonator, at which the maximum amplification is present*. That is, the eigenfrequency that lies closest to the frequency of the laser transition.

In many cases several eigenmodes of the resonator are amplified to a comparative degree such that the laser emits at many discrete frequencies. Since as a rule the frequency separation between the eigenmodes is very small, the frequency width of the laser radiation is still very small. Moreover, the different eigenmodes generally differ not only by their frequency, but rather by their intensity distribution over the cross section of the resonator. An important aim when constructing a

[8]"Mode" is short for "mode of propagation."

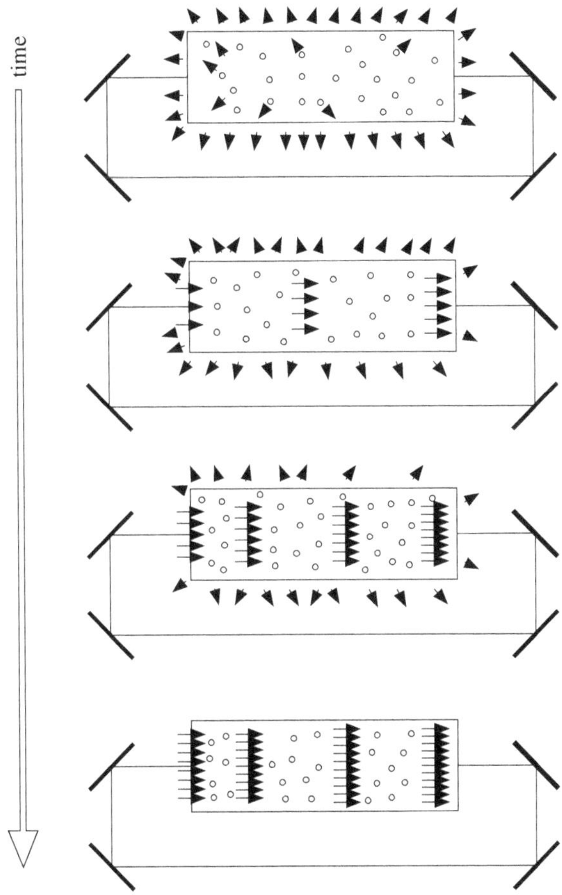

Fig. 2.12 Oscillation build-up of a laser. The spontaneous emission is suppressed little by little by the stimulated emission, since this is effectively fed back by the resonator

resonator is, thus, to attain the amplification of only one individual eigenmode; then the laser emits monochromatically[9] and in a very narrow spatial region.

The most important attribute of a resonator is its *sharpness of resonance*. The sharpness of resonance expresses how distinctive the resonance behavior is. A high resonance sharpness is reached with low resonator losses, since then the damping of the system is low and the amplification is extremely increased in the resonance case (Fig. 2.14). The frequency spectrum of the emitted radiation is essentially reflects the shape of the resonance curve. Therefore, the sharper the resonance, the narrower the frequency spectrum of the emitted radiation, and thus the closer one comes to the ideal monochromatic case. Through this, in turn, the *coherence* of the emitted light is increased. Indeed, the high degree of coherence belongs to one of the most important characteristics of laser light. The coherence and its significance will be discussed in the following section.

[9]Monochromatic: single-colored, which means only one light frequency appears.

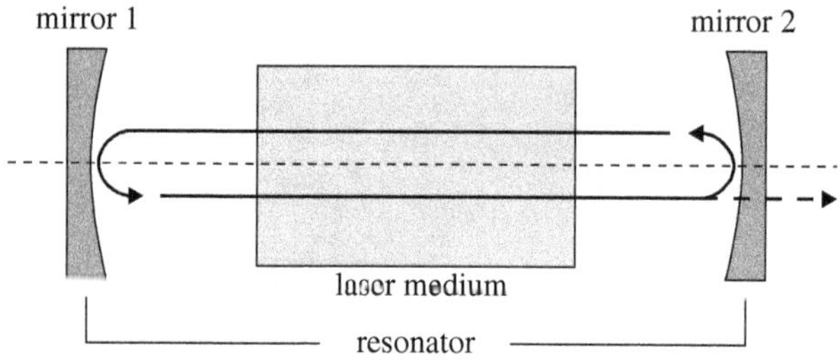

Fig. 2.13 Schematic representation of a laser resonator with imbedded laser medium. The radiation runs back and forth between the mirrors, a share of it is coupled out through mirror 2

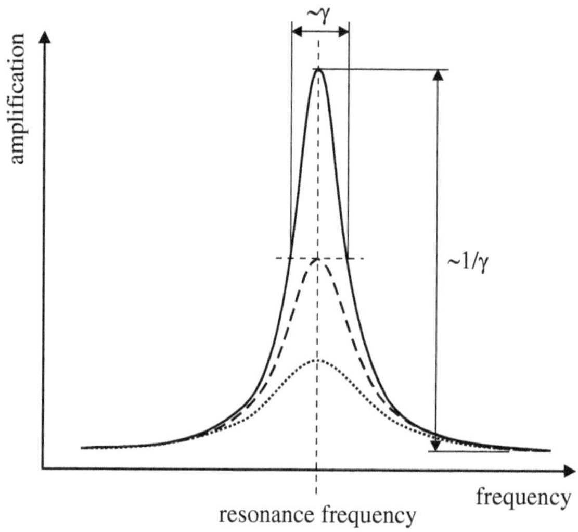

Fig. 2.14 The resonance curve describes the increase of the gain at the resonance frequency. The gain maximum is proportional to $1/\gamma$, when γ describes the damping caused by losses

2.5 Laser Radiation

2.5.1 Characteristic Properties

The great scientific and technological significance of the laser lies in the characteristic properties of laser radiation. They distinguish laser light from thermal light of conventional radiation sources. The most important characteristics have been listed below:

- *Very high intensities can be achieved with laser radiation.* The intensity is a measure of how much radiation power can be concentrated on a specific area. For many processes this represents one of the laser radiation's essential parameters.
- *Laser radiation is bundled and directed.* This is caused by the strict selection of direction through the resonator. A nearly parallel light beam results there, whose

widening is determined almost exclusively by the diffraction at the exit window of the resonator.[10] Directed beams can be focused to spots with very small diameters. On the one hand, this is important for laser-supported, spatially resolving measurement procedures, on the other, for numerous procedures of material processing. For the latter, it is important to deposit a very high amount of energy within a small area in a short period of time, because undesirable effects upon the surrounding material can thus be minimized.
- *Laser radiation is nearly monochromatic.* Above all, what is remarkable is that a laser can achieve both a small frequency width and a high intensity. Through spectral filters, a small frequency width can be selected as well, but only at the cost of the intensity. Monochromatic light of higher intensity is, for example, important for spectroscopic investigations.

The high monochromatism of laser radiation is equivalent to a *high coherency of the laser radiation.* The coherence is a measure of the length of continuous wave trains.[11] What this concept, continuous wave trains, means is that the electrical field oscillates without jumps in phase and significant fluctuations in amplitude. While conventional lamps produce wave trains of only several micrometers in length, the coherence length of lasers can reach several thousand kilometers. The coherence of radiation expresses itself in *interference phenomena.* These can, for example, be used to measure distances or speed with high precision.

- *Laser radiation exhibits an intensity statistic of narrow width.* Laser radiation is produced by a self-exciting oscillator. Such oscillators constantly emit an amplitude-stabilized signal, since amplitude fluctuations can be compensated by the strong nonlinear amplification and feedback. The consequence is that the intensity of the laser beam only fluctuates very marginally around a high mean value (*Poisson distribution*). Characteristic for thermal radiation is, in contrast, a wide intensity distribution, whose maximum constantly lies near zero (*Bose–Einstein distribution*). Therefore, the mean intensity of thermal light sources at the same peak intensity is much smaller than it is the case of laser radiation. There is no possibility of aligning the intensity statistic of thermal radiation with that of laser radiation.
- *Laser radiation can be produced in ultrashort pulses.* The shortest pulse length reached to date amounts to approximately 3 fs; the light pulse then encompasses only a few wave oscillations. When such short pulses are generated, the high coherence of laser radiation is exploited. Ultrashort pulses serve to enable temporally resolved investigation of plasma or solid-state characteristics.

The causes for these characteristics will be discussed in depth in the following chapters. Further examples of applications of the laser are presented in Sect. 2.6.

---

[10]The diffraction at the exit window determines the minimum possible expansion of the exiting beam; this is the so-called *diffraction limit*. Several modern lasers lie even at high output powers only a few percentage points above this limit.

[11]Here coherence means *longitudinal* or *temporal* coherence.

2.5.2 Laser Mode

The modes as resonant modes of oscillation of the laser resonator were introduced above, each of which is connected to a specific laser frequency. In this context, one speaks of *longitudinal modes*, since the characterization of the modes occurs via the field distribution along the resonator.

Since the electrical field has, however, to conform to specific boundary conditions perpendicularly to the propagation direction, different eigensolutions also result for the transverse field distribution, which are accordingly called *transverse modes*. Mostly transverse and longitudinal modes are not independent from each other in the strict sense.

The transverse modes of the resonator are important for two reasons. First, it is advantageous to have a field distribution inside the resonator, which optimally overlaps with the laser medium, in order to efficiently convert the pump power into laser radiation. Second, the transverse mode also determines the intensity distribution on the outcoupling mirror and thus in the beam cross section outside the resonator. The intensity distribution determined by the transverse resonator mode along the laser beam cross section is called the *laser mode*.

The essential systems of laser modes are the *Hermite-Gaussian modes* and the *Laguerre-Gaussian modes*. In Fig. 2.15 the intensity distributions of several modes of both the types are represented. The Hermite-Gaussian modes follow a Cartesian geometry, the Laguerre-Gaussian modes a cylindrical one. In both cases, the mode of the lowest order is the *Gaussian fundamental mode*. The significance of both Gaussian mode systems lies in that they are the transverse eigenmodes of a

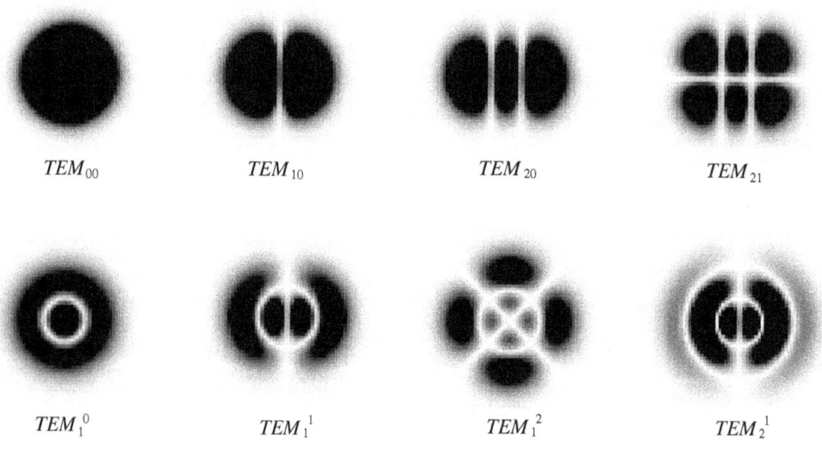

Fig. 2.15 Intensity distributions of several laser modes. In the upper row, Hermite-Gaussian modes, in the lower Laguerre-Gaussian modes. TEM stands for "Transverse electromagnetic mode." The indices indicate the order of the modes, equivalent with the number of the roots in the x/y direction, or in radial/azimuthal direction

resonator with spherically curved mirrors. This resonator configuration is the one almost exclusively encountered in practice.

2.5.3 Coherence

Radiation in the macroscopic sense—thus the emission of macroscopic radiation sources—arises through the succession of a large number of individual, atomic emission processes. Initially, the entirety of this atomic radiation results in the radiation field being emitted from the source. The properties of the radiation field arising from this superpositioning significantly depend, therefore, upon whether and which relationships there are between the individual, atomic emission processes.

In the case of thermal light sources, such as incandescent and discharge lamps, the spontaneous emission (cf. Sect 2.1.1) is the predominant process for producing light. Every individual emission event occurs spontaneously, this means *independent of the remaining emission events*. This independence can be interpreted in a spatial as well as in a temporal sense. The radiation of the individual atom is influenced neither by the behavior of the neighboring atom, nor by the events preceding temporally.

The "spontaneity" of the emission processes is transferred to the radiation released. Since the individual emission processes are not correlated, each of the released light quanta does not show any correlation either. What follows from this is that *the complete knowledge of the radiation field at a particular point in time and space does not allow any kind of statement about the radiation field at another point of time or another location*. Radiation of this kind is called spatially and temporally *incoherent*. A description of the radiation field is only possible using statistical mean values.

In contrast to thermal radiation sources, the light in a laser is produced almost exclusively by stimulated emission: an emission process stimulates further, exactly uniform processes. From this a strong, temporal as well as spatial dependency arises of each emission process from the others, so that the light quanta emitted are also strongly correlated. In this case the consequence is that *from the knowledge of the radiation field at one point in time and space, a very determined safe statement can be made about the radiation field at another point of time or another location*: this light is spatially and temporally coherent.

Inference experiments compare radiation that was emitted at different points of time or at different positions of a spatially expanded light source. With these experiments, spatial and temporal coherencies can thus be measured: in the case of coherent radiation, well-defined *inference* patterns result, whereas incoherent radiation only leads to a statistically distributed intensity. Coherence is a gradual property: radiation can be coherent to a smaller or larger extent. Temporal and spatial coherence can occur separately.

2.6 Fields of Applications of Laser Technology

The special properties of laser radiation make numerous scientific, technical, and industrial applications possible. As an example, Fig. 2.16 shows a selection from various technical sectors. Most of the examples named here have already been employed commercially. A significantly more comprehensive figure would result if all those sectors were also represented that are currently in research and development stages.

In the lower part of the figure, applications in communication and information technology predominate, which represent the laser's largest current field of use. Seen as its largest market segment of the future, optical computers and optical data processing are currently in, however, the research stage or only just being introduced to the market. An exception is the glass fiber cable, which has been used for long-distance connections in the telephone and data networks for many years.

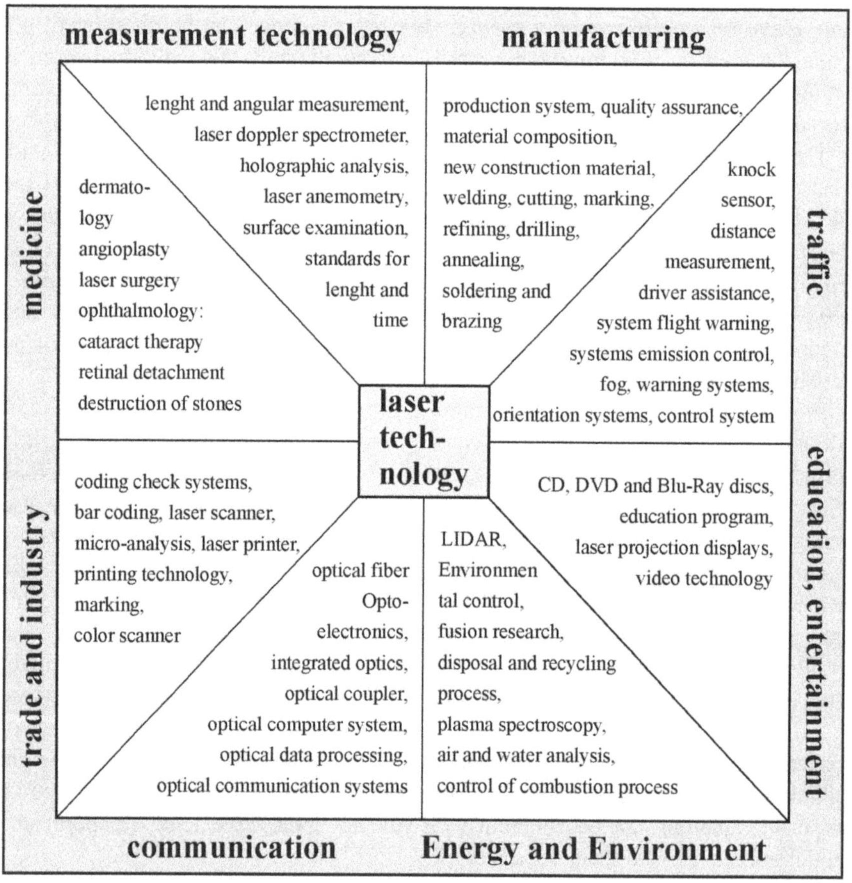

Fig. 2.16 Fields of application of laser technology

For many applications in the communication and information technology or office automation, the laser has set new benchmarks regarding quality and economy. Well-known examples are the tonal quality of the audio CD in comparison to a conventional LP record, as well as the precision and speed of laser printers, which are used today in most offices.

In medicine, lasers are being applied in diagnostics as well as in therapy. In both roles, the laser has proven itself and is used in several areas as a standard device, for example, in ophthalmology. In established procedures, doctors can follow and, if necessary, influence the laser's effects macroscopically and immediately by observing changes in the tissue as a result of coagulation, carbonization, vaporization, and ablation.

New therapeutic procedures are increasingly making use of processes running over reaction chains, as do photodynamic therapy in oncology or the use of microplasms in intraocular microsurgery. Typical for these applications is that the doctor can neither see nor access the influential process without an aid, and, therefore, on account of the high risk potential, a supplemental objective control procedure is required. Only using automatic control methods can doctors eliminate the risk of unintentionally damaging the healthy environment—since the processes, partially invisible, occur at high speed—in order to thus attain the best therapeutic benefit.

Laser technology has literally introduced new benchmarks into science and economy. In atomic and molecular physics, the laser has developed into an important resource, one which enables measurements with previously unattainable precision. Satellite geodesy and environmental controls through remote measurement of pollutants are now unthinkable without lasers. The precise and inexpensive measurement of lengths and coordinates, angles and velocities using a laser have become part and parcel of industrial measurement technology.

For modern manufacturing, random inspection no longer suffices. Much more, this sector is aiming at completely inspecting every part with correspondingly high requirements upon the speed of the testing procedures. For online measurement procedures with high processing speed, which can be integrated into the manufacturing process, laser technology offers excellent prerequisites, especially for methods of preventative quality assurance. As a contact-free measurement procedure, the laser-supported coordinate measurement technology is suitable for regulating manufacturing processes and for quality control in CIM and CAQ concepts.

The most important characteristics of processing with laser radiation can be summarized in the following points:

- A universal tool for the contact-free and wear-free processing and measuring,
- All working materials can be processed,
- High process speeds,
- Determining the treatment by selecting laser parameters, without retooling,
- High flexibility, and
- Optimal disposition for preventative methods of quality assurance.

The variety of possible applications demands a profound and foundational knowledge of how laser radiation and materials interact. Several methods of processing materials with laser radiation have been developed to such a degree that the user can apply the appropriate laser for the respective application according to the laser manufacturer's specification without any problems and with economical success. In these cases the machining process and its physical limits have been translated into technical rules. Currently this is limited to standard applications, such as cutting and welding of thin steel sheets as well as transformation hardening and remelting of cast iron. Research still has to be done to reach the thresholds of quality and efficiency during processing or to optimize the processing across the entire range of materials. Heat conductivity, melt pool dynamics, plasma formation, and the interaction of laser radiation with these processes have to be considered in each individual case.

Chapter 3
Electromagnetic Radiation

The theory of electromagnetic radiation belongs to the elementary foundations of laser technology. In this chapter, therefore, we will briefly outline the aspects of electromagnetic theory relevant to laser technology. For more detailed descriptions, we refer to the corresponding specialist literature.

3.1 The Spectrum of Electromagnetic Radiation

The spectrum of electromagnetic radiation significant to technical or scientific applications ranges from the lower frequency zone of technical alternating current ($\nu = 50$ Hz) to the gamma quanta of cosmic radiation ($\nu = 10^{22}$ Hz). In this zone, covering more than 20 orders of magnitude, the light a human eye can perceive covers a very small range ($\lambda = 0.4 - 0.75$ μm).

To classify the electromagnetic spectrum, three quantities (on different scales) illustrated in Fig. 3.1 are commonly used:

- The wavelength λ in the SI unit m (Meter),
- The frequency ν in the SI unit Hz (Hertz) or s^{-1}, and
- The energy E in the SI unit J (Joule). Energy is often indicated by the unit eV (electron volt).[1] In this context, the term *energy* refers to the energy of the elementary quanta of electromagnetic radiation, the photons:

$$E = h\nu$$

h Planck's Constant.

---

[1] 1 eV is the energy that an electron has acquired after passing through the potential difference of 1 V: 1 eV = $e \cdot 1$ V = 1.6×10^{-19} J, with the elementary charge $e = 1.6 \times 10^{-19}$ C.

R. Poprawe et al., *Tailored Light 1*, RWTHedition,
DOI 10.1007/978-3-642-01234-1_3

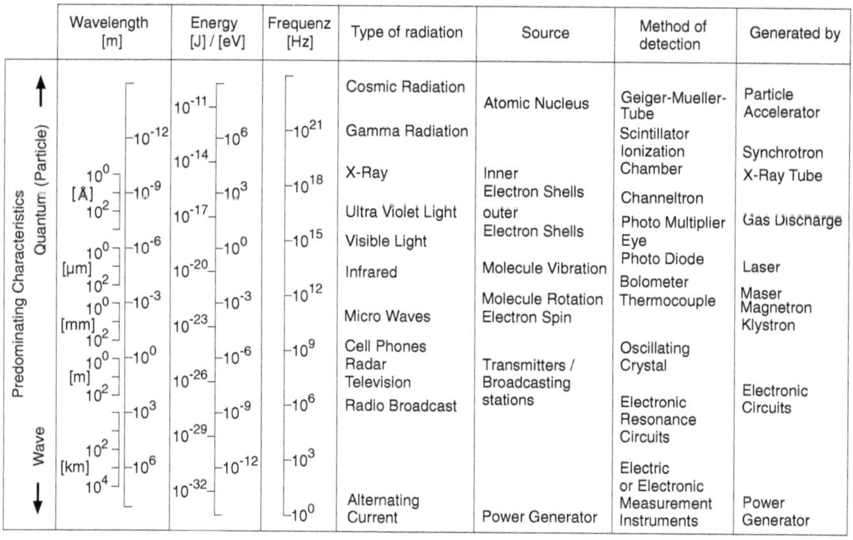

Fig. 3.1 The spectrum of electromagnetic radiation with examples for sources, detection, and technical generation

In addition, the wave number k in the unit m^{-1} is often used in spectroscopy. The wave number is proportional to the reciprocal value of the wave length:

$$k = \frac{1}{\lambda}.$$

Initially, the simultaneous existence of various scales had practical reasons. The experiments delivered wave numbers; the theories, however, frequencies or energies.

Due to

$$\lambda \nu = \frac{c_0}{n} \tag{3.1}$$

c_0 Speed of light in vacuum
n Refraction index of a medium

the conversion is simple in principle; for a long time, however, determining the speed of light c_0 was not possible with the required precision. Today, the different scales have a rather historic significance, since c_0 is known with sufficient precision, making the conversion simple.

In principle, there is no fixed wavelength range for laser radiation. Until now, the laser wavelengths used have been approximately between 0.1 and 700 μm; in this region, several thousand different laser wavelengths are known today. Thanks to continued development of laser technology, it can be expected that this number will increase and the range will expand further to the smaller wavelengths. The X-ray laser belongs to the important research areas in the sector of laser technology today.

In the long wavelength region, the spectrum of laser radiation borders on the maser range. In principle, the laser represents the transfer of the maser principle to smaller wavelengths, just as the maser can be seen as a high frequency realization of the feedback oscillator. The feedback oscillator then covers the entire frequency range up to the Hz region.

The laser's advance into the short wave X-ray region is of great significance for various applications. For example, the structures used in microelectronics are so small that they can no longer be imaged with visual light because the wavelength is too large. The X-ray laser could provide a remedy for this. Further important applications, for example in medicine or materials testing, are based on the characteristic of X-ray waves, which are able to deeply penetrate most materials and thus to image the interior of a body. Here, too, the X-ray laser would be a significant, new instrument. The construction of lasers in the X-ray region, however, has encountered the problem that the pump intensity necessary for generating population inversion increases proportionally to ν^3. In the short wavelength X-ray region, extremely high pump output is, thus, required.

Electromagnetic radiation in the usual laser wavelength regime is generated by electron transitions in the outer atom shells or by vibration and rotation transitions in molecules. Radiation of shorter wavelengths arises through electron transitions in the inner atom shells. Moreover, the continuous radiation of highly accelerated electrons can be used to generate laser radiation. This principle is applied in *free-electron lasers*.

The power of intensity of laser radiation can be measured according to two different principles: by measuring either the heating of a body due to impacting radiation (*thermoelement*) or the number of photons that hit a detector (*photodiode*, *photomultiplier*).

3.2 The Wave Equation

3.2.1 Maxwell's Equations

The *Maxwell Equations* represent the foundations of describing classic, electromagnetic phenomena. In particular, they establish the existence of electromagnetic waves in a vacuum. Formulated in 1873 by MAXWELL, they still belong to the fundamental basic equations of physics. Depending on how the problem is expressed, either the differential or the integral notation is used for Maxwell's Equations. For the theoretical description, the differential notation is generally more appropriate:

$$\begin{aligned}
&\vec{\nabla}\cdot\vec{D}=\rho \quad \text{Gaussian law for electric fields,}\\
&\vec{\nabla}\cdot\vec{B}=0 \quad \text{Gaussian law for magnetic fields,}\\
&\vec{\nabla}\times\vec{E}=-\frac{\partial\vec{B}}{\partial t} \quad \text{Faraday's law of induction,}\\
&\vec{\nabla}\times\vec{H}=\vec{j}+\frac{\partial\vec{D}}{\partial t} \quad \text{Ampère–Maxwell's law.}
\end{aligned} \tag{3.2}$$

With the fields

$\vec{E}$ Electric field strength,
$\vec{H}$ Magnetic field strength,
$\vec{D}$ Dielectric displacement, and
$\vec{B}$ Magnetic induction or flux density,

and

ρ Electric charge density (free charges),
$\vec{j}$ Current density (free currents).

The fields are related to each other via the *material equations*

$$\vec{D}(\vec{E})=\varepsilon_0\vec{E}+\vec{P}(\vec{E}),\quad \vec{B}(\vec{H})=\mu_0\left[\vec{H}+\vec{M}(\vec{H})\right] \quad \text{and} \quad \vec{j}=\vec{j}(\vec{E})\,. \tag{3.3}$$

$\varepsilon_0 = 8.86 \times 10^{-12}$ As/Vm Permittivity of free space,
$\mu_0 = 1.26 \times 10^{-6}$ Vs/Am Permeability of free space,
$\vec{P}$ Polarization of the medium,
$\vec{M}$ Magnetization of the medium.

The *Polarization* $\vec{P}$ and *Magnetization* $\vec{M}$ factor in the influence of the medium upon the field. In general, their dependency upon the field strengths can be non-linear. Polarization and magnetization can also be expressed by the polarization charge density ρ_{pol} and the magnetization current density j_{mag} (cf. Eqs. 3.7 and 3.8). Often *linearized material equations* are used instead of Eq. 3.3:

$$\vec{D}=\varepsilon_0\varepsilon\vec{E},\quad \vec{B}=\mu_0\mu\vec{H} \quad \text{and} \quad \vec{j}=\sigma\vec{E}\,. \tag{3.4}$$

ε Dielectric function,
μ magnetic susceptibility,
σ Electric conductivity.

The linearized relation between current density and electric field strength is described by *Ohm's Law*. For the field strengths occurring in classical optics, the linear approximations are generally valid. With lasers, significantly higher field strengths can be generated so that nonlinear effects also have to be considered; then terms of higher orders are added to the linear terms. In the context of problems in optics or in electrodynamics, ε, μ and σ are functions of the frequency. The direction of the polarization, or of the magnetization, does not have to initially be parallel to the respective field vector: therefore, ε, μ and σ are, generally speaking, *tensorial functions.*[2] In many cases they are, however, reduced to scalars if an isotropic medium is assumed.

3.2.2 The General Wave Equation

When used to describe the propagation of electromagnetic radiation, Maxwell's Equations are cumbersome in their basic form. For the sake of simplification, both of the last Maxwell's Equations, which describe the evolution of the electric and magnetic field in time, are often combined into the wave equation, while both of the first Maxwell's Equations deliver additional secondary conditions.

Applying the curl operator to the third Maxwell's Equation from Eq. 3.2, the following results

$$\vec{\nabla} \times \left(\vec{\nabla} \times \vec{E} \right) = -\frac{\partial}{\partial t} \vec{\nabla} \times \vec{B}.$$

On the left side, now the relation

$$\vec{\nabla} \times \vec{\nabla} \times \vec{E} = \vec{\nabla} \cdot \left(\vec{\nabla} \cdot \vec{E} \right) - \Delta \vec{E}$$

$$\Delta \vec{E} = \left(\frac{\partial^2}{\partial x^2} + \frac{\partial^2}{\partial y^2} + \frac{\partial^2}{\partial z^2} \right) \vec{E} \quad \text{Laplaceoperator in Cartesian coordinates}$$

is applied, and on the right side, the magnetic induction $\vec{B}$ is replaced by the magnetic field strength $\vec{H}$ using Eq. 3.3.

$$\vec{\nabla} \cdot \left(\vec{\nabla} \cdot \vec{E} \right) - \Delta \vec{E} = -\mu_0 \frac{\partial}{\partial t} \left(\vec{\nabla} \times \vec{H} + \vec{\nabla} \times \vec{M} \right). \tag{3.5}$$

Now, the magnetic field strength can be eliminated by inserting the fourth Maxwell Equation:

$$\vec{\nabla} \cdot \left(\vec{\nabla} \cdot \vec{E} \right) - \Delta \vec{E} = -\mu_0 \frac{\partial \vec{j}}{\partial t} - \mu_0 \varepsilon_0 \frac{\partial^2 \vec{E}}{\partial t^2} - \mu_0 \frac{\partial^2 \vec{P}}{\partial t^2} - \mu_0 \frac{\partial}{\partial t} \vec{\nabla} \times \vec{M}. \tag{3.6}$$

---

[2] ε and μ are then written as 3×3 matrices.

Thereby, a field relation from Eq. 3.3 was used, this time to express the dielectric displacement $\vec{D}$ by the electric field strength. Through the use of the first Maxwell's Equation and the definition of the *polarization charge density*, the first term on the left side can be rewritten

$$\rho_{\text{pol}} = -\vec{\nabla} \cdot \vec{P} \tag{3.7}$$

into

$$\vec{\nabla} \cdot \left(\vec{\nabla} \cdot \vec{E}\right) = \frac{1}{\varepsilon_0} \vec{\nabla} \cdot \left(\vec{\nabla} \cdot \vec{D} - \vec{\nabla} \cdot \vec{P}\right) = \frac{1}{\varepsilon_0} \vec{\nabla}\left(\rho + \rho_{\text{pol}}\right).$$

In this way, the definition of the polarization charge density is carried out analogous to the relation between the charge density ρ of free charges and the dielectric displacement in the first Maxwell's Equation. As an equivalent, the *magnetization current* is defined:

$$\vec{j}_{\text{mag}} = \vec{\nabla} \times \vec{M}. \tag{3.8}$$

Hereby, Eq. 3.6 becomes the *general wave equation of the electric field*

$$\Delta\vec{E} - \varepsilon_0\mu_0 \frac{\partial^2 \vec{E}}{\partial t^2} = \mu_0 \frac{\partial}{\partial t}\left(\vec{j} + \vec{j}_{\text{mag}}\right) + \frac{1}{\varepsilon_0}\vec{\nabla}\left(\rho + \rho_{\text{pol}}\right) + \mu_0 \frac{\partial^2}{\partial t^2}\vec{P}. \tag{3.9}$$

The general wave equation is an inhomogeneous, second order partial differential equation. It relates spatial and temporal changes of the electric field vector and permits the calculation of the propagation of electromagnetic waves at any point of space and time.

Only in rare cases is the complete wave equation, Eq. 3.9, applied to solve a problem. As a rule, the inhomogeneities on the right side can be completely or at least partially removed through appropriate assumptions. The following presents the most common variations for the wave equations derived in such a way.

An equivalent wave equation can also be derived for the magnetic induction $\vec{B}$. It is sufficient, however, to apply the wave equation for the electric field only; the magnetic field can then be determined by the Maxwell's Equations.

3.2.3 Wave Equation in Vacuum

The wave equation in vacuum represents the simplest, but also the most commonly used special case of the general wave equation. In a vacuum, free charges and free currents do not exist,

$$\rho = 0, \quad \vec{j} = 0\,.$$

In addition, the vacuum is assumed as unable to be polarized[3] and magnetized:

$$\vec{P} = 0 \Rightarrow \rho_{\mathrm{pol}} = 0, \quad \vec{M} = 0 \Rightarrow \vec{j}_{\mathrm{mag}} = 0.$$

In this way, all terms on the right side of Eq. 3.9 disappear, and the *wave equation for the electric field in vacuum* is

$$\Delta\vec{E} - \frac{1}{c_0^2}\frac{\partial^2\vec{E}}{\partial t^2} = 0, \quad \text{mit } c_0 = \frac{1}{\sqrt{\varepsilon_0\mu_0}}. \tag{3.10}$$

c_0 Speed of light in vacuum.

This wave equation is now a homogeneous, linear second order differential equation.

Historically, the discovery of wave propagation in free space represented a near revolution, since at this time many believed that wave propagation was always bound to a medium.[4]

3.2.4 *Wave Equations in Material*

The materials used in optical systems are, in general, not or only weakly able to be magnetized; mostly glasses, plastics, and various crystalline materials are used. It can also be assumed that free electric charges are not present, since these are nonmetallic materials. For the following derivations, it is hence assumed that

$$\vec{M} = 0 \Rightarrow \vec{j}_{\mathrm{mag}} = 0 \quad \text{or} \quad \mu = 1 \quad \text{and} \quad \rho = 0$$

is valid. For the connection between current density and electric field strength, Ohm's law is assumed to be valid (Eq. 3.4). If it is also assumed that the observed material is homogeneous and isotropic, then the conductivity is scalar and spatial as well as temporally constant. This assumption is only valid for electric field strengths that are not too large, because the isotropy of the medium will otherwise be destroyed by the electric field. Finally, it is also assumed, due to the homogeneity of the medium, that the polarization charge density is not dependent upon the location,

$$\nabla\rho_{\mathrm{pol}} = 0.$$

---

[3]This is only valid in classical theory, not in quantum electrodynamics. See also page XX.

[4]For a long time, the *ether theory* of light was adhered to: ether represented the propagation medium for light waves and was thought to fill the entire universe.

Under these assumptions, the wave equation, Eq. 3.9, becomes

$$\Delta\vec{E} - \mu_0\sigma\frac{\partial\vec{E}}{\partial t} - \frac{1}{c_0^2}\frac{\partial^2\vec{E}}{\partial t^2} = \mu_0\frac{\partial^2\vec{P}}{\partial t^2}. \tag{3.11}$$

This is the so-called *inhomogeneous wave equation*. The second term on the left side describes the propagation losses of the wave, which arise because the electric field strength is degraded by conduction currents in the medium.

The inhomogeneous term on the right side describes the influence of the wave propagation by the polarization of the medium, for example, a damping of the wave, or also an amplification if the medium exhibits an appropriate polarization. This equation is often chosen as starting point to describe the field in the laser medium. An amplifying effect of the polarization, however, is only possible if the expressions for the polarization are derived from a quantum mechanical model.

Under the further assumption that the polarization is, for small field strengths, linearly dependent upon the field strength, then the polarization can be expressed by the dielectric function ε or the electric susceptibility χ_s,

$$\vec{P} = \varepsilon_0\chi_e\vec{E} = \varepsilon_0(1-\varepsilon)\vec{E}, \tag{3.12}$$

χ_e Electrical susceptibility

whereas ε is a constant scalar due to the isotropy and homogeneity of the medium. Then the inhomogeneous wave equation can be simplified with

$$\frac{1}{c_0^2}\frac{\partial^2\vec{E}}{\partial t^2} - \mu_0\frac{\partial^2\vec{P}}{\partial t^2} = \frac{1}{c_0^2}\frac{\partial^2\vec{E}}{\partial t^2} - \frac{1}{\varepsilon_0 c_0^2}\varepsilon_0(1-\varepsilon)\frac{\partial^2\vec{E}}{\partial t^2} = \frac{\varepsilon}{c_0^2}\frac{\partial^2\vec{E}}{\partial t^2}$$

to the *homogeneous wave equation* or *the telegraph equation*

$$\Delta\vec{E} - \mu_0\sigma\frac{\partial\vec{E}}{\partial t} - \frac{1}{c^2}\frac{\partial^2\vec{E}}{\partial t^2} = 0. \tag{3.13}$$

$c = \frac{c_0}{n}$ Speed of light in the medium

$n = \sqrt{\varepsilon}$ Refraction index of the medium (Maxwell relation).

The telegraph equation is applied, above all, to describe electromagnetic waves in wave guides or dielectric optical fibers.

The assumption that the polarization is directly proportional to the field strength is only justified for field strengths and frequencies that are not too large. For very high frequencies, as they are in the optic range, the electric field of light waves and the polarization of the medium do not often behave synchronously, and for high field strengths, the polarization becomes dependent upon higher powers of the field strength. In these cases, the inhomogeneous equation cannot pass over to the homogeneous equation.

In nonconductive media, $\sigma = 0$, the damping term in Eq. 3.13 is also omitted, and the wave equation again takes on the same form as for the wave propagation in vacuum:

$$\Delta\vec{E} - \frac{1}{c^2}\frac{\partial^2\vec{E}}{\partial t^2} = 0. \tag{3.14}$$

This form of the wave equation is characterized as a *loss-free wave equation*. It is distinguished from the wave equation in vacuum by the propagation speed, which has changed by the factor of $1/n$.

Through an approach of the form

$$\vec{E} = \vec{E}_0 \sin \omega t \tag{3.15}$$

ω Angular frequency of the wave

the time derivation can subsequently be eliminated from the wave equation. What follows, then, is the time independent wave equation or the *Helmholtz equation*:

$$\Delta\vec{E}_0 + \frac{\omega^2}{c^2}\vec{E}_0 = 0. \tag{3.16}$$

In particular, the Helmholtz equation is used to describe diffraction phenomena or stationary field distribution in optical resonators.

3.2.5 Scalar Wave Equations

Often only the magnitudes of the fields, but not their orientation,[5] are important to describe a problem. In this case, it is less complicated to use scalar equations instead of vectorial wave equations.

If one observes, for example, the wave equation, Eq. 3.14, and breaks the field strengths down into their location and time-dependent contribution and a constant unit vector, the vector can be extracted from the equation,

$$\begin{aligned} \vec{E} = E \cdot \vec{e}: \quad & \Delta(E \cdot \vec{e}) - \frac{1}{c^2}\frac{\partial^2 (E \cdot \vec{e})}{\partial t^2} \\ & = \vec{e} \cdot \left(\Delta E - \frac{1}{c^2}\frac{\partial^2 E}{\partial t^2}\right) = 0 \Rightarrow \Delta E - \frac{1}{c^2}\frac{\partial^2 E}{\partial t^2} = 0, \end{aligned} \tag{3.17}$$

resulting in the scalar wave equation in Cartesian coordinates (Fig. 3.2).

[5]The *polarization* of the wave. The concept of polarization is introduced in Sect. 3.3.3.

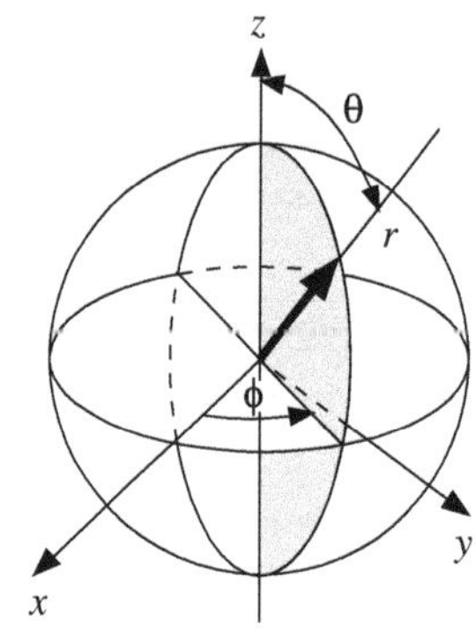

Fig. 3.2 Illustrating the spherical coordinates

In many cases, the scalar wave equation is also described in other coordinate systems, above all in *spherical coordinates*. With the Laplace operator in spherical coordinates, the scalar wave equation reads

$$E(r,\vartheta,\varphi,t):\quad \frac{1}{r}\frac{\partial^2(rE)}{\partial r^2}+\frac{1}{r^2\sin^2\vartheta}\left[\sin\vartheta\frac{\partial}{\partial\vartheta}\left(\sin\vartheta\frac{\partial E}{\partial\vartheta}\right)+\frac{\partial^2 E}{\partial\varphi^2}\right]-\frac{1}{c^2}\frac{\partial^2 E}{\partial t^2}=0. \tag{3.18}$$

It is very important to remember that the assumption is still valid that the vectorial part of the field strength; the vector $\vec{e}$, is spatially as well as temporally constant. Only then can a scalar wave equation be used. This means, however, that the individual components of $\vec{e}$ in spherical coordinates, e_r, e_ϑ and e_φ, are location dependent since the basic vectors of the spherical coordinates are location dependent on their own part. The significance of this point will become clearer when the solutions of the wave equation are treated in Sect. 3.3.4.

3.3 Elementary Solutions of the Wave Equation

The general solutions of the wave equations are extraordinarily varied. Only the specification of initial and boundary conditions limits their number. The fundamental characteristics of the general solutions can, however, be discussed by means of *elementary solutions* of the wave equations. From these elementary solutions, then more complex, general solutions can be formed through *superposition*.

3.3.1 Introduction to Complex Field Parameters

When oscillation and wave problems are solved, it is often advantageous to describe the fields by *complex functions*:

$$E = E_r + iE_i, \quad E^* = E_r - iE_i \quad \text{with } i = \sqrt{-1} \Leftrightarrow i^2 = -1. \tag{3.19}$$

E Complex field strength
E^* Conjugated complex field strength.

In particular, the exponential function is used with a complex argument in this context:

$$e^{i\varphi} = \cos\varphi + i\sin\varphi, \quad \left|e^{i\varphi}\right| = 1. \tag{3.20}$$

One of the advantages is that the two typical, real, and independent solutions for wave problems

$$E_+ \sim \cos, \quad E_- \sim \sin \tag{3.21}$$

are both contained in the complex solution. The extension to complex magnitudes, however, solely represents a mathematical resource. Only real magnitudes have physical significance. In the following, the physical, real field strength is defined as the real part of the complex field strength[6]:

$$\mathrm{E} = \frac{1}{2}(E + E^*) = E_r. \tag{3.22}$$

E Real field strength.

The magnetic induction and the remaining field parameters are approached in an equivalent way:

$$B = B_r + iB_i, \quad \mathrm{B} = \frac{1}{2}(B + B^*) = B_r \text{ etc.}$$

The real magnitudes can be gained from the complex magnitudes by adding the conjugated complex to it. For this the notation

$$\mathrm{E} = \frac{1}{2}(E + E^*) = \frac{1}{2}(E + \text{c.c.}) \tag{3.23}$$

c.c. Conjugated complex

is also used. Above all, when E is to be replaced by a complicated expression, this represents a practical shortcut.

Since the Maxwell Equations are linear in the field parameters, the transition from real to complex field parameters or vice versa is simple, as the example of Faraday's law of induction shows:

---

[6]Instead of this, the imaginary part could have also been chosen as a real field strength. These two possibilities reflect two independent, real solutions from Eq. 3.21.

$$\vec{\nabla} \times \vec{E} = -\frac{\partial \vec{B}}{\partial t}, \vec{\nabla} \times \vec{E}^* = -\frac{\partial \vec{B}^*}{\partial t} \Leftrightarrow \vec{\nabla} \times \left(\vec{E} + \vec{E}^*\right) \\ = -\frac{\partial\left(\vec{B} + \vec{B}^*\right)}{\partial t} \Leftrightarrow \vec{\nabla} \times \vec{\mathrm{E}} = -\frac{\partial \vec{\mathrm{B}}}{\partial t}. \tag{3.24}$$

Every equation for the complex fields can be conjugated, from which the corresponding relation results for the conjugated complex fields. From the sum of both equations, a valid relation follows for the real field parameters because of the linearity. This is also particularly valid for the wave equations derived from Maxwell's Equations.

3.3.2 Planar Waves

According to the problem involved, it can be practical to formulate the wave equation in different coordinate systems. The most important are the Cartesian coordinates, cylindrical coordinates, spherical coordinates, and the general spheroid coordinates, which include the Cartesian and the spheroid coordinates as a limit case. Since the solutions in the different coordinate systems are solutions of the same equation in a different representation, they can always be translated into each other.

Due to the fundamental character of its solutions, the wave equation, Eq. 3.14, is solved in Cartesian coordinates. The ansatz

$$\vec{E}_{\vec{k}} = \vec{E}_{0,\vec{k}} \mathrm{e}^{i\vec{k}\vec{x} - i\omega t} \tag{3.25}$$

$\vec{k}$ Wave vector

fulfills the differential equation, when the *dispersion relation* between k and ω

$$k^2 = \frac{\omega^2}{c^2}, \quad k = \left|\vec{k}\right| = \frac{2\pi}{\lambda} \tag{3.26}$$

k Wave number,
c Speed of light in the propagation medium,

is valid. The solutions in Eq. 3.25 are characterized as *planar waves*, since points of the same phase, respectively, lie on planes at a distance of a wavelength. The wave vector indicates the propagation orientation and forms the normal vector on the phase planes. This requirement results directly from Maxwell's Equations, when the charge density ρ disappears:

$$\vec{\nabla} \cdot \vec{E} = \vec{k} \cdot \vec{E} = 0 \Rightarrow \vec{k} \perp \vec{E}. \tag{3.27}$$

For the magnetic induction, the following

$$\vec{\nabla} \cdot \vec{B} = \vec{k} \cdot \vec{B} = 0 \Rightarrow \vec{k} \perp \vec{B} \tag{3.28}$$

is valid as an equivalent. Waves of this kind are named *transversal waves.*[7] Due to the third Maxwell Equation (Faraday's law of induction), the following relation between $\vec{E}$ and $\vec{B}$ is valid for planar waves:

$$\vec{k} \times \vec{E} = \omega \vec{B} \Rightarrow \vec{E} \perp \vec{B} \quad \text{and} \quad \left|\vec{E}\right| = \frac{\omega}{k}\left|\vec{B}\right| = c \cdot \left|\vec{B}\right|. \tag{3.29}$$

The speed of light c is the *phase speed* of the wave, e.g., the speed at which the planes of constant phase move through space:

$$\vec{k}\vec{x} - \omega t = \text{const.} \Rightarrow \vec{v}_{\text{Ph}} \equiv \frac{\mathrm{d}\vec{x}}{\mathrm{d}t} = \frac{\vec{k}}{k}\frac{\omega}{k} = \vec{e}_k \frac{\omega}{k}, \; \left|\vec{v}_{\text{Ph}}\right| = \frac{\omega}{k} = c. \tag{3.30}$$

Due to the linearity of the wave equation, every superposition of elementary solutions is a solution. The general solution of the wave equation reads, therefore (Fig. 3.3),

$$\vec{E} = \sum_{\vec{k}} \vec{E}_{\vec{k}} = \sum_{\vec{k}} \vec{E}_{0,\vec{k}} \mathrm{e}^{i\vec{k}\vec{x} - i\omega t}. \tag{3.31}$$

Any solution of the wave equation can be represented as the sum of planar waves,[8] which means that the planar waves form a complete system of elementary solutions. The individual planar wave is always only an idealized approximate solution in the physical sense: since its amplitude remains constant on the planes of the same phase, the planar wave has an infinite extension in the x, y, and z directions.

3.3.3 Polarization of Electromagnetic Waves

Due to Eqs. 3.27 and 3.28, the field vectors are perpendicular to the propagation direction. The direction of the vectors has, thus, not yet been determined. Only when the *polarization* of the wave is indicated are the field vectors finally defined.

---

[7]Purely transversal waves only exist in uncharged, unlimited space. The presence of charges on limiting surfaces always induces longitudinal components.

[8]This is then the so-called *Fourier representation* of the solution.

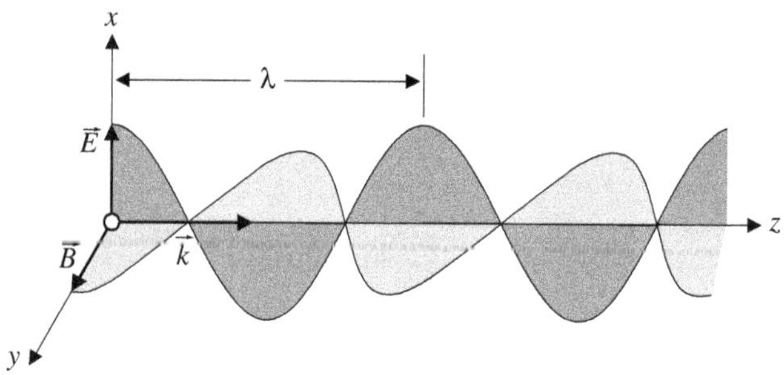

Fig. 3.3 The transversal, planar electromagnetic wave. Electric field, magnetic field, and propagation orientation are perpendicular to each other

If the coordinate axes are chosen such that the wave propagates along the z-axis, then the electric field can have components along the x-axis and the y-axis:

$$\vec{k} = \begin{pmatrix} 0 \\ 0 \\ k \end{pmatrix} \Rightarrow \vec{E} = \begin{pmatrix} E_x \\ E_y \\ 0 \end{pmatrix} = \begin{pmatrix} E_{0,x} \\ E_{0,y}\mathrm{e}^{i\delta} \\ 0 \end{pmatrix} \mathrm{e}^{ikz-i\omega t} \tag{3.32}$$

In addition to different magnitudes, the x and y components can still differ by a phase difference δ. To define the polarization, it is common to combine the components of the E field vector that are orthogonal to the propagation direction into a two-component, normalized vector—the *polarization vector* or *Jones vector*.

$$\vec{e} = \frac{1}{\sqrt{E_{0,x}^2 + E_{0,y}^2}} \begin{pmatrix} E_{0,x} \\ E_{0,y}\mathrm{e}^{i\delta} \end{pmatrix}, \quad |\vec{e}| = 1. \tag{3.33}$$

By means of the polarization vector, the three foundational polarization forms of the wave can be distinguished from each other.

Linear Polarization

Linear polarization is present when the phase difference is

$$\delta = m\pi, \quad m = 0, \pm 1, \pm 2, \ldots. \tag{3.34}$$

It is characterized by the electric field vector retaining its oscillation direction. The plane spanned by the vectors $\vec{E}$ and $\vec{k}$ is called the *polarization plane*. There are two polarization directions, orthogonal to each other, which can be chosen along the Cartesian axes:

$$\vec{e}_x = \begin{pmatrix} 1 \\ 0 \end{pmatrix}, \quad \vec{e}_y = \begin{pmatrix} 0 \\ 1 \end{pmatrix}. \tag{3.35}$$

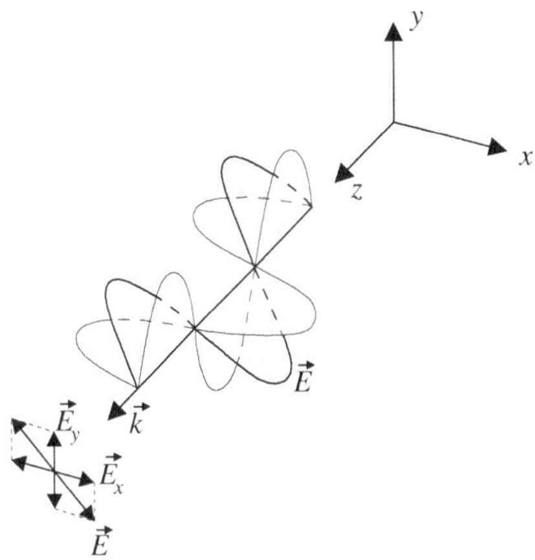

Fig. 3.4 Linearly polarized planar wave. The field vector oscillates in a plane

Through superpositioning, all other polarization directions can be formed from the two independent polarization vectors (Fig. 3.4).

Circular Polarization
For circular polarization the phase difference is

$$\delta = \frac{2m+1}{2}\pi, \quad m = 0, \pm 1, \pm 2, \dots \tag{3.36}$$

and it is $E_{0,x} = E_{0,y}$. The polarization vector is then

$$\vec{e}_{r,l} = \frac{1}{\sqrt{2}}\begin{pmatrix} 1 \\ \pm i \end{pmatrix}. \tag{3.37}$$

In this case, the direction of the field vector is not fixed. If, namely, the temporal dependency of the wave is added,

$$\vec{e}_{r,l} \cdot \mathrm{e}^{-i\omega t} = \frac{1}{\sqrt{2}}\begin{pmatrix} \mathrm{e}^{-i\omega t} \\ \pm i.\mathrm{e}^{-i\omega t} \end{pmatrix} = \frac{1}{\sqrt{2}}\begin{pmatrix} \cos\omega t - i\sin\omega t \\ \pm\sin\omega t \pm i\cos\omega t \end{pmatrix}, \tag{3.38}$$

then it becomes clear that the field vector rotates to the left or the right according to the sign of the second component of the polarization vector, whereas its magnitude remains constant, however, due to $E_{0,x} = E_{0,y}$: its head describes a circle in the x-y plane (Fig. 3.5).

The case of the upper leading sign is called right-circular polarization, that of the lower sign left-circular polarization:

$$\text{Right-circular polarization: } \delta = +\frac{\pi}{2} + 2m\pi, \quad m = 0, \pm 1, \pm 2, \dots, \text{ and}$$

$$\text{Left-circular polarization: } \delta = -\frac{\pi}{2} + 2m\pi, \quad m = 0, \pm 1, \pm 2, \dots.$$

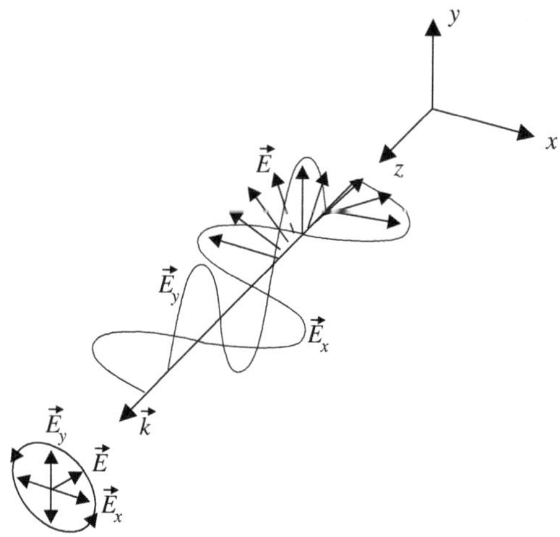

Fig. 3.5 Circularly polarized planar wave. The field vector describes a turning movement around the propagation axis

These are again two independent polarization vectors from whose superposition every other polarization form can be constructed (Fig. 3.6).

Elliptical Polarization

The elliptical polarization is the most general polarization state. The phase difference δ as well as the amplitude relation of $E_{0,x}$ to $E_{0,y}$ is arbitrary. The linear and circular polarization are, therefore, included as special cases. The electrical field vector rotates, as for the circular polarization, around the propagation direction, whereas this time, its head describes an ellipsis, however, due to $E_{0,x} \neq E_{0,y}$. The axial ratio of the ellipse is given by the amplitude relation. Every elliptical polarization state can be produced by superposing a circularly polarized and a linearly polarized state (Fig. 3.7).

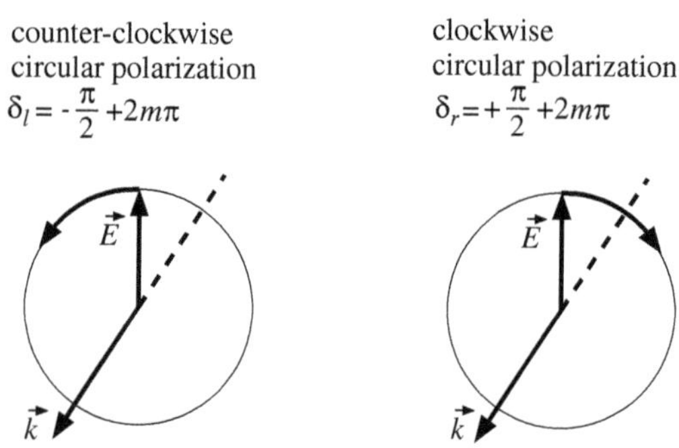

Fig. 3.6 Left-circular and right-circular polarization

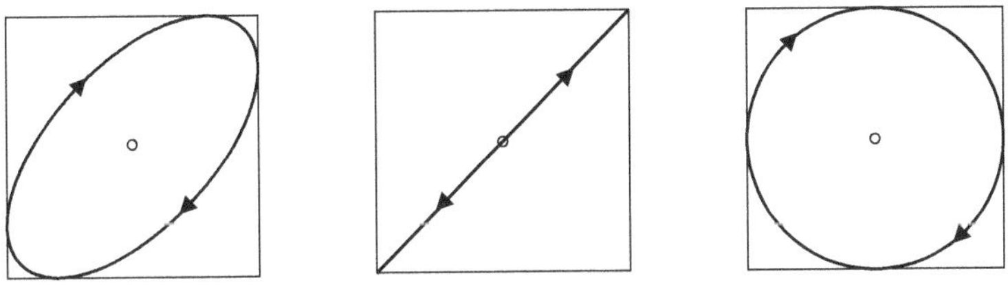

Elliptic Linear Circular

Fig. 3.7 Movement of the head of the field vector around the propagation axis

Polarizers and Phase Shifters

Natural light is unpolarized, which means it contains a noncorrelated mixture of all polarization states. By using polarizers and phase shifters, the individual polarization directions can be filtered out and transformed into each other.

Polarizers are made of conductive lines that lie close and in parallel. For microwaves and infrared radiation, wires tensioned in parallel can be used; for radiation of shorter wavelengths, the polarizer consists of a thin film in which long-chained, conductive molecules are embedded. The polarizer only lets through the component of the incident radiation whose electric field vector is perpendicular to the conductive lines. In this way, a certain, linear polarization direction can be filtered out of the radiation.

Phase Shifters are thin discs made of a birefringent material, which means the material exhibits a different refraction index for different, linear polarization directions. Due to the different refraction index, one polarization direction propagates quicker than the other. Through the appropriate choice of the thickness of the disc, a specific phase difference can be set between the components of the electric field vector.

In the case of the *$\lambda/4$ phase plate*, a phase difference of $\pi/2$ is attained. The comparison with the phase conditions in Eqs. 3.34 and 3.36 show that this corresponds to the conversion of linearly polarized radiation into circularly polarized radiation[9] and vice versa.

The *$\lambda/2$ phase plate*, in contrast, leads to a phase difference of π. It does not have a visible influence upon linearly polarized light. This slab does, however, reverse the rotational direction of circularly or elliptically polarized light.

[9]In general, this results in elliptically polarized radiation. Only with a determined orientation of the $\lambda/4$ slab to the polarization level is the resulting radiation circularly polarized: the electric field vector has to be divided into exactly the same parts on both refraction directions of the slab.

3.3.4 Spherical Waves

The planar waves turn out to be solutions of the vectorial wave equation in Cartesian coordinates (x, y, z). In addition to these, the solution in *spherical coordinates* (r, φ, ϑ) is, above all, relevant.

For the vectorial wave equation, it is difficult to determine the solutions in the spherical coordinates. The reason for this is that the vectorial wave equation does not have spherical-symmetrical solutions: it is impossible to arrange tangential vectors upon a spherical surface in such a way that no vortices or poles result.

There are, however, two arguments that support searching for spherical-symmetrical solutions of the *scalar* wave equation, thus while neglecting the direction of the field vector:

- At least over large ranges of the sphere's surface, the field vectors can be arranged without vortices, and over a certain expansion, their direction in space can be seen as parallel. Then, the transition to a scalar wave equation is acceptable mathematically (cf. Sect. 3.2.5). In this sense, the spherically symmetrical solution of the scalar wave equation represents an approximation for the wave propagation in a limited angular range.
- Natural light sources emit, as a rule, a static mixture of waves of all polarization directions in every direction. At every location, the polarization direction of the field changes, therefore, in a statistical progression. In this case, an explicit description of the polarization is not practical, and one can limit oneself to the scalar wave equation.

The scalar wave equation in spherical coordinates is (cf. Eq. 3.18)

$$\frac{1}{r}\frac{\partial^2(rE)}{\partial r^2} + \frac{1}{r^2\sin^2\vartheta}\left[\sin\vartheta\frac{\partial}{\partial\vartheta}\left(\sin\vartheta\frac{\partial E}{\partial\vartheta}\right) + \frac{\partial^2 E}{\partial\varphi^2}\right] - \frac{1}{c^2}\frac{\partial^2 E}{\partial t^2} = 0. \tag{3.39}$$

It is solved by means of the separation ansatz

$$E(r,\varphi,\vartheta,t) = E_0 \cdot u(r) \cdot g(\varphi,\vartheta) \cdot \mathrm{e}^{-i\omega t}. \tag{3.40}$$

The solutions for the radial part $u(r)$ and the angular part $g(\varphi, \vartheta)$ are then given by special functions:

$$\begin{array}{l} u_1(r) = k \cdot j_l(kr), \quad u_2(r) = k \cdot n_l(kr) \\ g(\varphi,\vartheta) = Y_l(\vartheta) \end{array} \quad \text{with} \quad k = \tfrac{\omega}{c}, \quad l = 0,\ 1,\ 2, \ldots. \tag{3.41}$$

The functions Y_n, which determine the angular part of the spherical waves, are the *spherical functions*:

$$\begin{aligned} Y_n(\vartheta) &\equiv Y_n^{m=0}(\varphi,\vartheta) = \sqrt{\tfrac{2n+1}{4\pi}}P_n(\cos\vartheta), \\ P_n(z) &= \tfrac{1}{2^n n!}\tfrac{d^n}{dz^n}(z^2-1)^n, \qquad n = 0, 1, 2, \ldots. \end{aligned} \tag{3.42}$$

The polynomials $P_n(z)$, which occur in them, are the so-called *Legendre Polynomials*. Therewith the first three spherical functions are

$$Y_0(\vartheta) = \frac{1}{4\pi}, \quad Y_1(\vartheta) = \sqrt{\frac{3}{4\pi}}\cos\vartheta, \quad Y_2(\vartheta) = \sqrt{\frac{5}{4\pi}}\left(\frac{3}{2}\cos^2\vartheta - \frac{1}{2}\right). \tag{3.43}$$

For the radial part, there are at first two different kinds of solutions. The solutions $u_1(r)$ *regular* at the zero point are given by the *spherical Bessel functions* $j_l(z)$, which can be derived in turn from the Bessel functions $J_{n+1/2}$ with half integral order,

$$j_l(z) = \sqrt{\frac{\pi}{2z}}J_{n+\frac{1}{2}}(z), \tag{3.44}$$

or they result from the following differential representation:

$$\begin{aligned} j_l(z) = (-z)^l\left(\frac{1}{z}\frac{\mathrm{d}}{\mathrm{d}z}\right)^l\frac{\sin z}{z} \Rightarrow j_0(z) &= \frac{\sin z}{z}, \\ j_1(z) &= \frac{\sin z}{z^2} - \frac{\cos z}{z}, \ldots \end{aligned} \tag{3.45}$$

The solutions *singular* at $r = 0$ are the *spherical Neumann functions* $n_l(z)$:

$$\begin{aligned} n_l(z) &= (-1)^{l+1}\sqrt{\frac{\pi}{2z}}J_{-l-\frac{1}{2}}(z) \\ &= -(-z)^l\left(\frac{1}{z}\frac{\mathrm{d}}{\mathrm{d}z}\right)^l\frac{\cos z}{z} \Rightarrow n_0(z) = -\frac{\cos z}{z} \\ &\qquad\qquad n_1(z) = -\frac{\cos z}{z^2} - \frac{\sin z}{z}, \ldots. \end{aligned} \tag{3.46}$$

If the argument approaches zero, the functions can thus be simplified to

$$\begin{aligned} z \to 0: j_l(z) &\approx \frac{z^l}{(2l+1)!!}, \quad n_l(z) \approx \frac{-(2l-1)!!}{z^{l+1}}, \\ (2l+1)!! &= 1\cdot 3\cdot 5\cdot \ldots \cdot(2l+1), \end{aligned} \tag{3.47}$$

which means the spherical Bessel functions remain finite at $z = 0$, whereas the spherical Neumann functions diverge there. For very large z, on the other hand, the following is valid

$$z \to \infty: j_l(z) \approx \frac{1}{z}\sin\left(z - \frac{l\pi}{2}\right), \quad n_l(z) \approx -\frac{1}{z}\cos\left(z - \frac{l\pi}{2}\right), \tag{3.48}$$

where the functions in essence thus decline with $1/z$. Apart from a phase shift, their form is independent of the order l.

As a rule, the singular and the regular solutions are combined in the complex notation:

$$\begin{aligned} u_l^+(r) &= ikj_l(kr) - kn_l(kr) = \frac{e^{ikr - il\frac{\pi}{2}}}{r} \\ u_l^-(r) &= -ikj_l(kr) - kn_l(kr) = \frac{e^{-ikr + il\frac{\pi}{2}}}{r} \end{aligned} \quad \text{for } kr \gg 1 \Leftrightarrow r \gg \tfrac{\lambda}{2\pi}. \tag{3.49}$$

The complete solutions of the lowest order

$$l = 0: E(r, \vartheta, t) = E_0 u_0^{\pm}(r) Y_0(\vartheta) e^{-i\omega t} = \frac{E_0}{4\pi}\frac{e^{\pm ikr - i\omega t}}{r} = \tilde{E}_0 \frac{e^{\pm ikr - i\omega t}}{r} \tag{3.50}$$

are characterized as *spherical waves*, since for these solutions the surfaces of constant phases lie on concentric spherical surfaces at every point of time $t = t_0$:

$$t = t_0 : \quad \pm kr - \omega t_0 = \text{const.} \Rightarrow r = \pm\frac{\omega}{k}t_0 + \text{const.} = \pm ct_0 + \text{const.} \tag{3.51}$$

According to the leading sign, the wave expands to smaller or larger radii: the two possible solutions describe *incoming* and *outgoing* spherical waves.

Since the solutions of the wave equation can be represented, on the one hand, by planar waves, as shown in the last section, and on the other, by the solutions of the representation in the spherical coordinates presented here, both kinds of solutions have to be convertible into each other. In fact, the following relation can be derived between the planar waves, on the one hand, and the spherical Bessel functions and the spherical functions, on the other:

$$e^{i\vec{k}\vec{r}} = e^{ikr\cos\vartheta} = \sum_{l=0}^{\infty} i^l (2l+1) j_l(kr) P_l(\cos\vartheta), \tag{3.52}$$

whereas the spherical functions were reduced directly upon the Legendre polynomials. The planar wave can be represented as a superposition of the spherical waves.

Praxis suggests one always choose that coordinate system to solve the wave equation in which the phase surfaces have the simpler form. Except in Cartesian and spherical coordinates, solutions of the wave equations are known in numerous other systems; these two are, at least in classical optics, the ones most commonly used: as a rule, light that is emitted from a source can be described using outgoing spherical waves; at a great distance or after corresponding focussing by a lens, by approximation planar waves are present (Fig. 3.8).

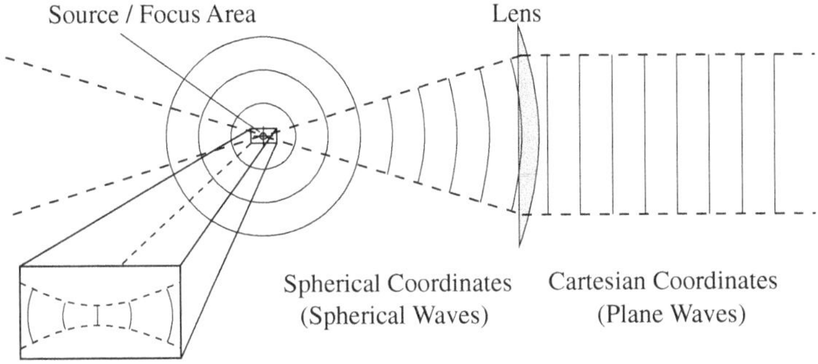

Fig. 3.8 Transformation of spherical waves in planar waves (or vice versa) by a lens. The focal area, or the immediate surrounding of the source, cannot be described by spherical waves. For this, Gaussian-Hermite solutions have to be used. (cf. equation, Chap. 5)

Difficulties arise in the description of the light waves very close to a *focal point*. There the description using spherical waves leads to nonphysical results, since the solution formulated in Eq. 3.50 then diverges: the field strength would be infinite in the focal point. A way out of this problem is offered by the description using *spheroid coordinates*, or by the *Gaussian-Hermite solutions* for electromagnetic waves. There it can be seen that the focus cannot become a point, but rather remains limited to a diameter larger than the wavelength. For large distances from the focal area, the spheroid solutions merge into the spherical waves. For the strongly directed and precisely focussable laser radiation, the description of the focal area is very important. Therefore, this kind of focussed radiation is dealt in Chap. 5.

3.3.5 *Energy Density of Electromagnetic Waves*

An energy conservation law can be derived from the Maxwell Equations for the real fields by multiplying the third equation (Faraday's Law of Induction, cf. Eq. 3.2) with $\vec{\mathrm{H}}$ and the fourth equation (Ampere-Maxwell's Law) with $\vec{\mathrm{E}}$:

$$\begin{aligned} \vec{\mathrm{H}}\cdot\left(\vec{\nabla}\times\vec{\mathrm{E}}\right)+\vec{\mathrm{H}}\cdot\tfrac{\partial\vec{\mathrm{B}}}{\partial t}&=0\\ \vec{\mathrm{E}}\cdot\left(\vec{\nabla}\times\vec{\mathrm{H}}\right)-\vec{\mathrm{E}}\cdot\tfrac{\partial\vec{\mathrm{D}}}{\partial t}&=\vec{j}\cdot\vec{\mathrm{E}}. \end{aligned} \quad (3.53)$$

The real fields have to be used here, since Eq. 3.53 is no longer linear and, therefore, the expansion to complex functions cannot be performed as in Sect. 3.3.1. The dielectric displacement $\vec{\mathrm{D}}$ and the magnetic induction $\vec{\mathrm{B}}$ are expressed by the field strengths using the relations in linear approximation according to Eq. 3.3:

$$\vec{\mathrm{D}} = \varepsilon_0 \varepsilon \cdot \vec{\mathrm{E}}, \quad \vec{\mathrm{B}} = \mu_0 \mu \cdot \vec{\mathrm{H}}. \tag{3.54}$$

If one subtracts the second equation from the first in Eq. 3.53, the following results

$$\frac{1}{2}\frac{\partial}{\partial t}\left(\mu_0 \mu \vec{\mathrm{H}}^2 + \varepsilon_0 \varepsilon \vec{\mathrm{E}}^2\right) + \vec{\nabla} \cdot \left(\vec{\mathrm{E}} \times \vec{\mathrm{H}}\right) = -\vec{j} \cdot \vec{\mathrm{E}} - \frac{1}{2}\left(\mu_0 \vec{\mathrm{H}}^2 \frac{\partial \mu}{\partial t} + \varepsilon_0 \vec{\mathrm{E}}^2 \frac{\partial \varepsilon}{\partial t}\right), \tag{3.55}$$

whereas the relationships

$$\begin{gathered} \vec{\mathrm{H}} \frac{\partial}{\partial t}\left(\mu \vec{\mathrm{H}}\right) = \frac{1}{2}\frac{\partial}{\partial t}\left(\mu \vec{\mathrm{H}}^2\right) + \frac{1}{2}\vec{H}^2 \frac{\partial \mu}{\partial t} \quad \text{and} \\ \vec{\nabla} \cdot \left(\vec{E} \times \vec{\mathrm{H}}\right) = \vec{\mathrm{H}} \cdot \left(\vec{\nabla} \times \vec{\mathrm{E}}\right) - \vec{\mathrm{E}} \cdot \left(\vec{\nabla} \times \vec{\mathrm{H}}\right) \end{gathered} \tag{3.56}$$

were used.[10] Using the definitions

$w_{em} = \frac{1}{2}\left(\varepsilon_0 \varepsilon \vec{E}^2 + \mu_0 \mu \vec{\mathrm{H}}^2\right)$ Energy density of the electromagnetic field
$\vec{S} = \vec{E} \times \vec{\mathrm{H}}$ Energy current density (Poynting vector)
$p_{mech} = \vec{j} \cdot \vec{E}$ Mechanical power density

the general principle of the conservation of energy for electromagnetic waves originates:

$$\frac{\partial}{\partial t} w_{em} + \vec{\nabla} \cdot \vec{S} = -p_{mech} - \frac{1}{2}\left(\mu_0 \vec{\mathrm{H}}^2 \frac{\partial \mu}{\partial t} + \varepsilon_0 \vec{E}^2 \frac{\partial \varepsilon}{\partial t}\right). \tag{3.57}$$

If ε and μ are not temporally dependent, then the so-called *Poynting theorem* of the principle of the conservation of energy follows:

$$\frac{\partial}{\partial t} w_{em} + \vec{\nabla} \cdot \vec{S} = -p_{mech}. \tag{3.58}$$

The Poynting vector is perpendicular on both field vectors; in the case of a planar wave, it points, therefore, in the direction of the wave vector $\vec{k}$. Hence, the second term describes the transport of energy with the propagation of the wave. The term on the right side expresses the transformation of energy in mechanical power, if currents are evoked by the electric fields. The terms appearing additionally in

[10]Here it was assumed that ε and μ are scalar. In the tensorial case, the expressions in Eqs. 3.55 and 3.56 have to be written as quadratic forms, but in principle nothing changes due to this.

Eq. 3.57 account for the energy required to change the magnetization, or the polarization, of the wave medium, expressed by the magnetic susceptibility μ and the dielectric function ε. In general, these contributions can be neglected.

The energy current density, averaged over time and over a wave period T, is called radiation strength or *intensity* of the radiation:

$$I = \left\langle |\vec{S}| \right\rangle = \langle |\vec{E} \times \vec{\mathrm{H}}| \rangle, \quad \left\langle |\vec{S}| \right\rangle \equiv \frac{1}{T} \int_0^T |\vec{S}|\, dt. \tag{3.59}$$

$T = \frac{2\pi}{\omega} = \frac{1}{\nu}$ Wave period.

For purely transversal waves, the expressions for energy density, energy current density and intensity can be simplified further. Due to the relationship of Eq. 3.29 between electrical field strength and magnetic induction, the magnetic field strength $\vec{\mathrm{H}}$ can be expressed by $\vec{E}$ and hence eliminated. Therefore, energy density, the Poynting vector and intensity for transversal waves are given by

$$w_{\mathrm{em}} = \varepsilon_0 \varepsilon \mathrm{E}^2, \quad \vec{S} = c w_{\mathrm{em}} \frac{\vec{k}}{k} \quad \text{and} \quad I = \left\langle |\vec{S}| \right\rangle = c \langle w_{\mathrm{em}} \rangle = \frac{\varepsilon_0 \varepsilon c}{2} \mathrm{E}_0^2. \tag{3.60}$$

E_0 Amplitude of the electric field.

For the time averaging, it was assumed that the electric field has a harmic time dependence:

$$\vec{E} = \vec{E}_0 \cos \omega t \Rightarrow \langle \mathrm{E}^2 \rangle = \mathrm{E}_0^2 \frac{1}{2\pi/\omega} \int_0^{2\pi/\omega} \mathrm{d}t \cos^2 \omega t = \frac{1}{2} \mathrm{E}_0^2. \tag{3.61}$$

The field energy density $\langle w_{\mathrm{em}} \rangle$, averaged over a wave period, is also called the *wave energy density*.

Since all freely propagating, electromagnetic waves are transverse, the simplified Eq. 3.60 is more commonly used than the more general definitions according to Eq. 3.56. Nontransverse waves appear above all in wave guides and plasmas; there the simplified relations are, therefore, invalid.

Primarily in the technical sector, the prefactors appearing in Eq. 3.60 are often expressed by the *wave impedance* Z of the medium:

$$Z = \sqrt{\frac{\mu \mu_0}{\varepsilon \varepsilon_0}} = \mu \mu_0 c = \frac{1}{\varepsilon \varepsilon_0 c} \Rightarrow \vec{S} = \frac{\mathrm{E}^2}{Z} \frac{\vec{k}}{k}, I = \frac{1}{2} \frac{\mathrm{E}_0^2}{Z}. \tag{3.62}$$

Z Wave impedance.

The wave impedance of the vacuum is

$$Z_0 = \sqrt{\frac{\mu_0}{\varepsilon_0}} = 376.72\,\Omega. \tag{3.63}$$

The wave impedance is especially significant when an electromagnetic wave should be coupled from one medium into another: the wave is reflected and the coupling efficiency becomes smaller proportionally to the difference between the wave impedances of the media.

Finally, attention is drawn to a relation between the complex and the time-averaged, real field strengths, which is often used. As a rule, when electromagnetic waves are described, expressions are initially available for the complex field strengths. The dependency upon time is mostly harmonic:

$$\vec{E} \sim \vec{H} \sim \mathrm{e}^{-i\omega t}, \quad \Rightarrow \vec{E}^* \sim \vec{H}^* \sim \mathrm{e}^{i\omega t}. \tag{3.64}$$

According to Eq. 3.22 this holds:

$$\vec{E} = \frac{1}{2}\left(\vec{E}+\vec{E}^*\right) = \Re\left(\vec{E}\right), \quad \mathrm{H} = \frac{1}{2}\left(\vec{H}+\vec{H}^*\right) = \Re\left(\vec{H}\right). \tag{3.65}$$

The time average of a product of two real field magnitudes can be directly expressed, therefore, by the complex magnitudes in the following way:

$$\begin{aligned} \langle \vec{E}\vec{H} \rangle &= \frac{1}{4}\langle (\vec{E}+\vec{E}^*)(\vec{H}+\vec{H}^*) \rangle \\ &= \frac{1}{4}\langle \vec{E}\vec{H}+\vec{E}^*\vec{H}^*+\vec{E}^*\vec{H}+\vec{E}\vec{H}^* \rangle = \frac{1}{4}\left(\vec{E}^*\vec{H}+\vec{E}\vec{H}^*\right) = \frac{1}{2}\Re\left(\vec{E}^*\vec{H}\right), \end{aligned} \tag{3.66}$$

due to

$$\langle \vec{E}\vec{H} \rangle \sim \tfrac{\omega}{2\pi}\int_0^{2\pi/\omega} \mathrm{d}t\, \mathrm{e}^{-2i\omega t} = 0, \qquad \langle \vec{E}^*\vec{H}^* \rangle \sim \tfrac{\omega}{2\pi}\int_0^{2\pi/\omega} \mathrm{d}t\, \mathrm{e}^{2i\omega t} = 0,$$

$$\langle \vec{E}^*\vec{H} \rangle \sim \langle \vec{E}\vec{H}^* \rangle \sim \tfrac{\omega}{2\pi}\int_0^{2\pi/\omega} \mathrm{d}t\,1 = 1.$$

For the square of a field magnitude, in the same way

$$\left\langle \vec{E}^2 \right\rangle = \frac{1}{2}\vec{E}^*\vec{E} = \frac{1}{2}\left|\vec{E}\right|^2 \tag{3.67}$$

follows, which means the time averaging can be replaced by the square of the absolute value of the complex field strength. With this, the intensity from Eq. 3.60 becomes

$$I = c\langle w_{\text{em}}\rangle = c\varepsilon_0\varepsilon\left\langle \vec{E}^2 \right\rangle = \frac{c\varepsilon_0\varepsilon}{2}\left|\vec{E}\right|^2 = \frac{1}{2Z}\left|\vec{E}\right|^2 \tag{3.68}$$

3.4 Superposition of Waves

In Sect. 3.3, it was already indicated that due to the linearity of the wave equation each superposition of the elementary solutions also represents a solution to the wave equation. The superposition of wave solutions leads to a number of different superposition phenomena. For this, the result depends upon the parameter by which the superposing waves differ.

Again, using planar waves for the discussion appears most expedient here. Each planar wave is well-defined by the following parameters:

1. The frequency ν, or the magnitude of the wave vector k,
2. The propagation direction, given through the direction of the $\vec{k}$-vector,
3. The phase of the wave δ,
4. The amplitude $\left|\vec{E}\right|$, and
5. The polarization, represented by the polarization vector $\vec{e}$.

The amplitude only plays a subordinate role in the superpositioning processes. Four different, fundamental superpositioning processes can be correspondingly distinguished.

3.4.1 *Superposition with Different Phases*

In the simplest and, at the same time, fundamental case, the superposing waves differ only by a relative phase difference δ:

$$\vec{E}_1 = \vec{E}_0 \mathrm{e}^{ikz - i\omega t}, \quad \vec{E}_2 = \vec{E}_0 \mathrm{e}^{ikz - i\omega t}\mathrm{e}^{i\delta}. \tag{3.69}$$

For simplicity's sake, the propagation was assumed to occur along the z-axis. From the superposition, the following wave then results

$$\begin{aligned} \vec{E}_{12} &= \vec{E}_1 + \vec{E}_2 = \vec{E}_0 \mathrm{e}^{ikz - i\omega t}\left(1 + \mathrm{e}^{i\delta}\right), \\ \left|\vec{E}_{12}\right| &= \left|\vec{E}_0\right| \cdot \left|1 + \mathrm{e}^{i\delta}\right| = \sqrt{2(1 + \cos\delta)} \cdot \left|\vec{E}_0\right|. \end{aligned} \tag{3.70}$$

The process of superpositioning waves of the same frequency and propagation direction is called *interference*, the total wave arising from this the *interference wave*. The amplitude of the interference wave now depends upon the relative phase

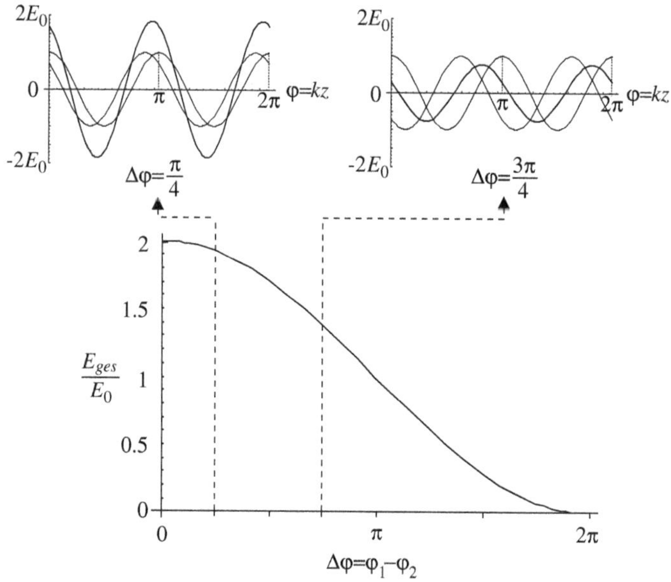

Fig. 3.9 The total field strength for the interference of two waves, depending upon the phase difference $\Delta\varphi$

of the individual waves. Most notably, the cases of *constructive* and *destructive interference* are of interest:

$$\begin{aligned}&\text{constructive interference: } \delta = 2m\cdot\pi \Rightarrow \left|\vec{E}_{12}\right| = 2\left|\vec{E}_0\right| \\ &\text{destructive interference: } \delta = (2m+1)\cdot\pi \Rightarrow \left|\vec{E}_{12}\right| = 0\end{aligned} \quad m = 0, \pm1, \pm2, \ldots \tag{3.71}$$

In the case of constructive interference, the waves superposition exactly in phase, and the total amplitude is the sum of the individual amplitudes. For the destructive interference, the waves superposition exactly out of phase and extinguish each other (Fig. 3.9).

The interference of waves with shifted phase is used in interferometers, as in the *Michelson Interferometer*, to determine marginal differences in path lengths, which in turn are, e.g., caused by uneven or not completely parallel surfaces in optical systems.

3.4.2 Superposition of Differently Polarized Waves

In Sect. 3.3.3, it was already mentioned that only two independent polarization states are needed to construct the other states. For this, either the two orthogonal, linear polarization vectors,

$$\vec{e}_x = \begin{pmatrix} 1 \\ 0 \end{pmatrix}, \quad \vec{e}_y = \begin{pmatrix} 0 \\ 1 \end{pmatrix}, \tag{3.72}$$

or the two opposed turning vectors of the circular polarization states,

$$\vec{e}_r = \frac{1}{\sqrt{2}} \begin{pmatrix} 1 \\ +i \end{pmatrix}, \quad \vec{e}_l = \frac{1}{\sqrt{2}} \begin{pmatrix} 1 \\ -i \end{pmatrix}, \tag{3.73}$$

present themselves as a basis. In order to determine the polarization vector resulting from the superposition of two waves, the polarization vectors of the waves, multiplied by the amplitudes and relative phases, have to be added and subsequently normalized.

Thus, a linearly polarized wave can, for example, be produced by superposing a left- and right-circular wave:

$$\frac{1}{\sqrt{2}} [\vec{e}_r + \vec{e}_l] = \begin{pmatrix} 1 \\ 0 \end{pmatrix} = \vec{e}_x. \tag{3.74}$$

If the circularly polarized waves are superposed with a phase difference of π, the other polarization direction then results:

$$\frac{1}{\sqrt{2}} \left[\vec{e}_r + e^{i\pi}\vec{e}_l\right] = \frac{1}{\sqrt{2}} [\vec{e}_r - \vec{e}_l] = \begin{pmatrix} 0 \\ i \end{pmatrix} = i\vec{e}_y = e^{i\frac{\pi}{2}}\vec{e}_y. \tag{3.75}$$

In the same way, both circular polarizations can be formed out of linearly polarized waves, if both waves are superpositioned with a phase difference of $\pi/2$ or $-\pi/2$:

$$\frac{1}{\sqrt{2}} \left[\vec{e}_x + e^{i\frac{\pi}{2}}\vec{e}_y\right] = \frac{1}{\sqrt{2}} \begin{pmatrix} 1 \\ i \end{pmatrix} = \vec{e}_r \quad \text{and} \quad \frac{1}{\sqrt{2}} \left[\vec{e}_x + e^{-i\frac{\pi}{2}}\vec{e}_y\right] = \frac{1}{\sqrt{2}} \begin{pmatrix} 1 \\ -i \end{pmatrix} = \vec{e}_l. \tag{3.76}$$

Correspondingly, one can imagine every polarized wave as the superposition of two other orthogonal polarizations. This idea is advantageous when understanding the passage a wave takes through an arrangement of polarizers and phase shifters. The incident wave is dispersed into parts parallel and perpendicular to the polarizer's orientation. Only the perpendicular part is transmitted.

3.4.3 *Superposition of Waves of Different Frequency*

The superposition of waves of different frequency produces an appearance widely known from sound waves: the beat. If both waves are given by

$$\vec{E}_1 = \vec{E}_0 \mathrm{e}^{ik_1 z - i\omega_1 t} \quad \text{and} \quad \vec{E}_2 = \vec{E}_0 \mathrm{e}^{ik_2 z - i\omega_2 t} \tag{3.77}$$

then from their superposition, it follows that

$$\vec{E}_{12} = \vec{E}_1 + \vec{E}_2 = \vec{E}_0 \left(\mathrm{e}^{ik_1 z - i\omega_1 t} + \mathrm{e}^{ik_2 z - i\omega_2 t}\right) = 2\vec{E}_0 \cos(k_- z - \omega_- t)\, \mathrm{e}^{ik_+ z - i\omega_+ t}. \tag{3.78}$$

$\omega_- = \frac{\omega_1 - \omega_2}{2}$, $k_- = \frac{k_1 - k_2}{2}$ Modulation frequency and wave number
$\omega_+ = \frac{\omega_1 + \omega_2}{2}$, $k_+ = \frac{k_1 + k_2}{2}$ Mean frequency and wave number

Here, it was used that

$$\cos\alpha = \frac{1}{2}\left(\mathrm{e}^{i\alpha} + \mathrm{e}^{-i\alpha}\right)$$

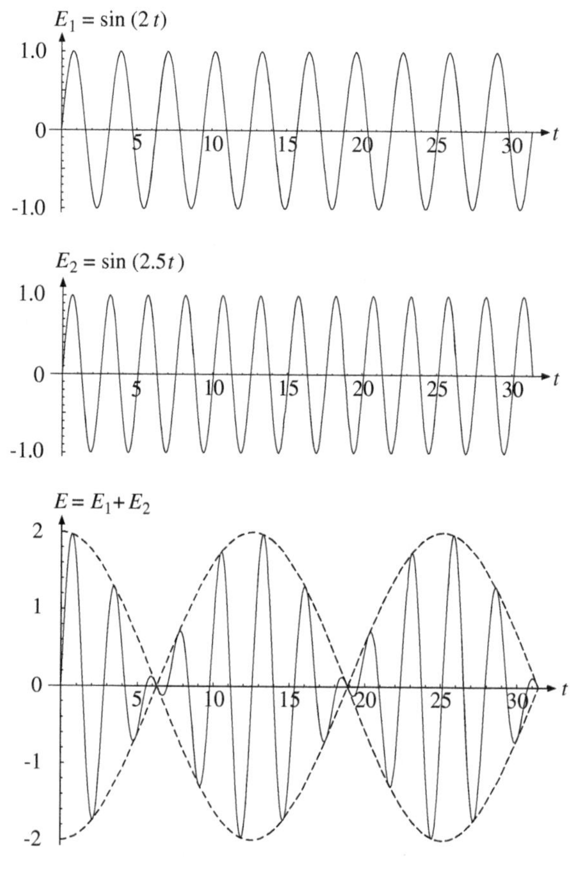

Fig. 3.10 The beat of two sinus waves at the frequencies $\omega_1 = 2$ and $\omega_2 = 2.5$

holds. The sum of the two waves can be written as the product of two oscillation terms, such that a wave with the *mean frequency* now appears, which is modulated with a lower frequency, the *modulation frequency* (Fig. 3.10). The closer both output frequencies are to each other, the lower the modulation frequency is.

The intensity of a wave is proportional to the square of the field strength,

$$I \sim \left|\vec{E}_{12}\right|^2 = 4E_0^2 \cos^2(k_- z - \omega_- t) = 2E_0^2[1 + \cos(2k_- z - 2\omega_- t)], \tag{3.79}$$

which means it is modulated with the double modulation frequency. This modulation of the intensity is called the *beat*; the double modulation frequency is the *oscillation frequency*.

3.4.4 Group Velocity and Dispersion

The superposition of two waves of different frequency is the simplest example for a *wave packet*: in general, wave packets comprise a number of waves of different frequencies with a defined phase relation. In the case of the two waves, as in the last section, each wave antinode of the modulated wave corresponds to a wave packet.

The wave antinodes move with the *group velocity* along the propagation direction. It results, therefore, from the condition

$$k_- z - \omega_- t = \text{const.} \Rightarrow v_{gr} = \frac{\mathrm{d}z}{\mathrm{d}t} = \frac{\omega_-}{k_-}, \tag{3.80}$$

or more generally, when the wave packet is composed of continuous frequency spectrum:

$$\begin{aligned} \omega_2 \to \omega_1 + \mathrm{d}\omega,\ k_2 \to k_1 + \mathrm{d}k \Rightarrow \omega_- &\to 2\mathrm{d}\omega,\ k_- \to 2\mathrm{d}k \\ \Rightarrow v_{gr} &= \frac{\mathrm{d}\omega}{\mathrm{d}k}. \end{aligned} \tag{3.81}$$

v_{gr} Group velocity.

In a vacuum $\omega = c_0 k$ holds, so that the group velocity corresponds to the speed of light in a vacuum. In many media, the refraction index n is dependent upon the frequency; the group velocity also varies with the frequency:

$$c = \frac{c_0}{n} = \frac{\omega}{k_0 n} = \frac{\omega}{k} \Rightarrow k = k_0 n \Rightarrow v_{gr} = \frac{\mathrm{d}\omega}{\mathrm{d}k} = \frac{1}{k_0}\frac{\mathrm{d}\omega}{\mathrm{d}n} = \frac{1}{k_0}\left(\frac{\mathrm{d}n}{\mathrm{d}\omega}\right)^{-1} \tag{3.82}$$

c_0, k_0 Speed of light and wave number in vacuum,
n Refraction index of the medium,
c, k Speed of light and wave number in the medium.

This dependency is called *dispersion*; the function $n(\omega)$ is the *dispersion relation.*[11] In vacuum, no dispersion is present. Dispersion expresses itself primarily in wave packets being deformed as they pass through the medium, since the parts with higher frequency propagate with a different velocity than those with a lower frequency.

3.4.5 Superposition of Waves with Different Propagation Directions

If two planar waves of the same frequency and polarization, however with different propagation directions, overlap, then the following image results. Both waves are given by

$$\vec{E}_1 = \vec{E}_0 e^{ik\vec{e}_1\vec{x}-i\omega t} \quad \text{and} \quad \vec{E}_2 = \vec{E}_0 e^{ik\vec{e}_2\vec{x}-i\omega t} \tag{3.83}$$

$\vec{e}_1, \vec{e}_2$ Unit vectors in propagation direction of the first or second wave.

The coordinate system is chosen in such a way that both propagation vectors lie in the x-z plane and are symmetrically inclined to the z-axis by the angle $\pm\varphi$ (Fig. 3.11):

$$\vec{e}_1 = \begin{pmatrix} -\sin\varphi \\ 0 \\ \cos\varphi \end{pmatrix}, \quad \vec{e}_2 = \begin{pmatrix} \sin\varphi \\ 0 \\ \cos\varphi \end{pmatrix}. \tag{3.84}$$

The superposition of both waves leads to

$$\begin{aligned} \vec{E} = \vec{E}_1 + \vec{E}_2 &= \vec{E}_0 e^{ikz\cos\varphi}\left(e^{ikx\sin\varphi} + e^{-ikx\sin\varphi}\right)e^{-i\omega t} \\ &= 2\vec{E}_0 \cos(kx\sin\varphi)e^{ikz\cos\varphi - i\omega t} \\ &= 2\vec{E}_0 \cos(\kappa x)e^{i\tilde{k}z - i\omega t}, \\ \text{with} \quad & \kappa = k\sin\varphi,\ \tilde{k} = k\cos\varphi. \end{aligned} \tag{3.85}$$

This is a wave propagating in the z direction with reduced wave number $\tilde{k}$, by the factor cos (φ), which means the larger the angle 2φ is, which is enclosed by both partial waves, the smaller the wave number of the total wave is. In the x direction, the amplitude of the total wave is modulated according to cos (κx); the wave number of the modulation κ becomes larger, the larger the enclosed angle is.

In the limit case $\varphi = \pi/2$, the partial waves propagate along the x-axis, but in opposing directions. For the superposition, in this case the following results

[11] In general, all relations between the frequency ω and the wave number k are called dispersion relations; thereby, the wave number in the medium is proportional to the refraction index.

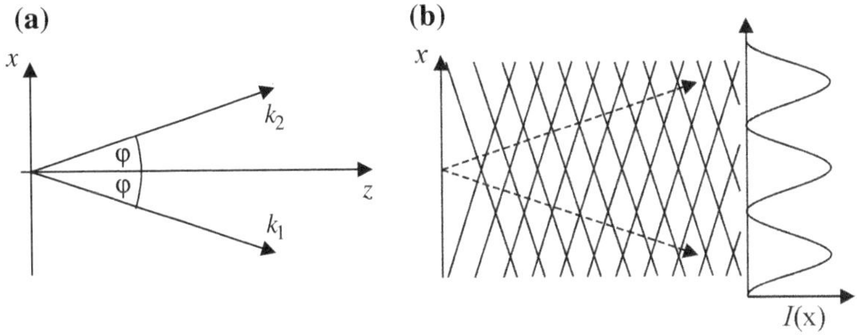

Fig. 3.11 **a** The choice of the coordinate system. **b** The overlapping of the phase fronts and the resulting interference pattern

$$\varphi = \frac{\pi}{2} \Rightarrow \tilde{k} = 0, \kappa = k, \quad \vec{E} = 2\vec{E}_0 \cos(kx) e^{-i\omega t}. \tag{3.86}$$

At the locations where

$$kx = (2n+1)\frac{\pi}{2}, \quad n = 0, \pm 1, \pm 2, \ldots \tag{3.87}$$

holds, the field strength disappears at any point of time (*wave nodes*); while in between it exhibits a temporally independent, complex amplitude and a phase oscillating with the frequency ω. Due to the stationary wave node, this wave type is called a *standing wave*. Since the Poynting vectors of the two partial waves are of the same magnitude and directed in opposition to each other, the Poynting vector of the total wave is zero,

$$\vec{S} = \vec{S}_1 + \vec{S}_2 = 0, \tag{3.88}$$

which means *the standing wave does not transport any energy*. There is, however, field energy stored in it:

$$w_{\mathrm{em}} = \varepsilon\varepsilon_0 \mathrm{E}^2 = 2\varepsilon\varepsilon_0 |E_0|^2 \cos^2(kx) \cos^2(\omega t). \tag{3.89}$$

The intensity is modulated corresponding to the stationary wave nodes,

$$I \equiv I(x) = c\langle w_{\mathrm{em}} \rangle = c\varepsilon\varepsilon_0 |E_0|^2 \cos^2 kx. \tag{3.90}$$

This stationary modulation of the intensity can actually be measured. For microwaves, by measuring the distance between the minimum intensities, the wavelength of the radiation can be identified.

Chapter 4
The Propagation of Electromagnetic Waves

In the previous chapter, electromagnetic waves were introduced as a fundamental phenomenon. All electromagnetic radiation fields can be described using the equations and elementary solutions presented there. The general and exact solution is, however, either too complex or cannot be determined at all. Therefore, the special characteristics of each problem will be used to simplify the description.

In the case of the laser, electromagnetic radiation propagates primarily tightly concentrated and in a defined direction, in the so-called *radiation direction*: the laser emits a "light beam." In both of the following chapters, different and idealized models of light beams are presented in principle. Each of these models is adapted to specific situations and valid, therefore, only under specific assumptions.

Lasers are not the only devices that emit light beams. Thus, the models presented here are just as valid for "classic" light beams as they are for laser beams. It will be seen, however, that the real laser beam comes closest to the idealized theoretical description.

In general, the theories which describe the propagation of light beams are united under the concept of *Optics*.

4.1 Propagation Regimes and Fresnel Number

As mentioned already, a directed and confined light beam can be produced without using a laser, for example, by illuminating a circular opening in an aperture with a lamp (Fig. 4.1).

By using a spectral lamp or an appropriate filter, the beam can also be made mostly monochromatic. If the light is then caught with a screen behind the aperture, the intensity distribution of the light shows a different behavior depending upon the distance to the aperture. It proves to be practical to define the dimensionless number N_F, the *Fresnel number*, as a parameter for the distance to the aperture.

R. Poprawe et al., *Tailored Light 1*, RWTHedition,
DOI 10.1007/978-3-642-01234-1_4

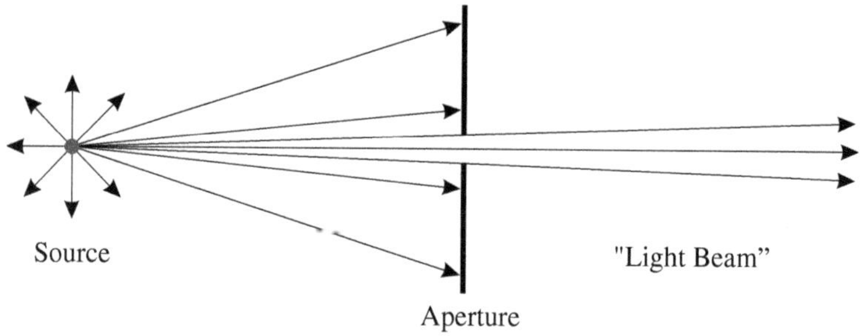

Fig. 4.1 Producing a "light beam" by illuminating an aperture

$$N_F = \frac{r^2}{z\lambda} \tag{4.1}$$

r Radius of the aperture
z Distance from the observation plane to the aperture
λ Wave length of the radiation

By means of the Fresnel number, three regions are distinguished (Fig. 4.2). Directly behind the aperture, the Fresnel number is very large. The intensity distribution of the light mirrors the geometry of the opening exactly, which means the shadow borders are sharp and the intensity is constant over the beam cross-section. The light only propagates along geometrical lines towards the light source. This region is called, therefore, the region of *geometrical optics.*

If, in contrast, the distance to the aperture is very large, then the Fresnel number is close to zero. This is the region of the *Fraunhofer diffraction*. The sharp shadow lines become softer and the beam broadens over the geometrical shadow lines. The intensity distribution over the beam's cross section has a smooth contour, decreasing toward the boundary of the beam. While the width of the intensity distribution grows with growing distance to the aperture, its form remains constant. For a circular aperture, the opening cone of the beam is given by the angle

$$\sin\theta \cong 1.22\frac{\lambda}{2r}. \tag{4.2}$$

which is called the *Fraunhofer diffraction angle*, or shortened simply to diffraction angle. Due to the blurring of the shadow lines, details of the aperture geometry are no longer visible from large distances. More and more, the intensity distributions behind apertures of different shapes align themselves with the distribution behind a circular aperture, until each opening appears to be point-shaped in infinite distance.

The transition region between large and small Fresnel numbers is called the region of *Fresnel diffraction*. At this point the Fresnel number is of the magnitude of 1. Fundamentally, the radiation still propagates within the geometrical shadow lines; the intensity distribution, however, is strongly structured and quickly changes

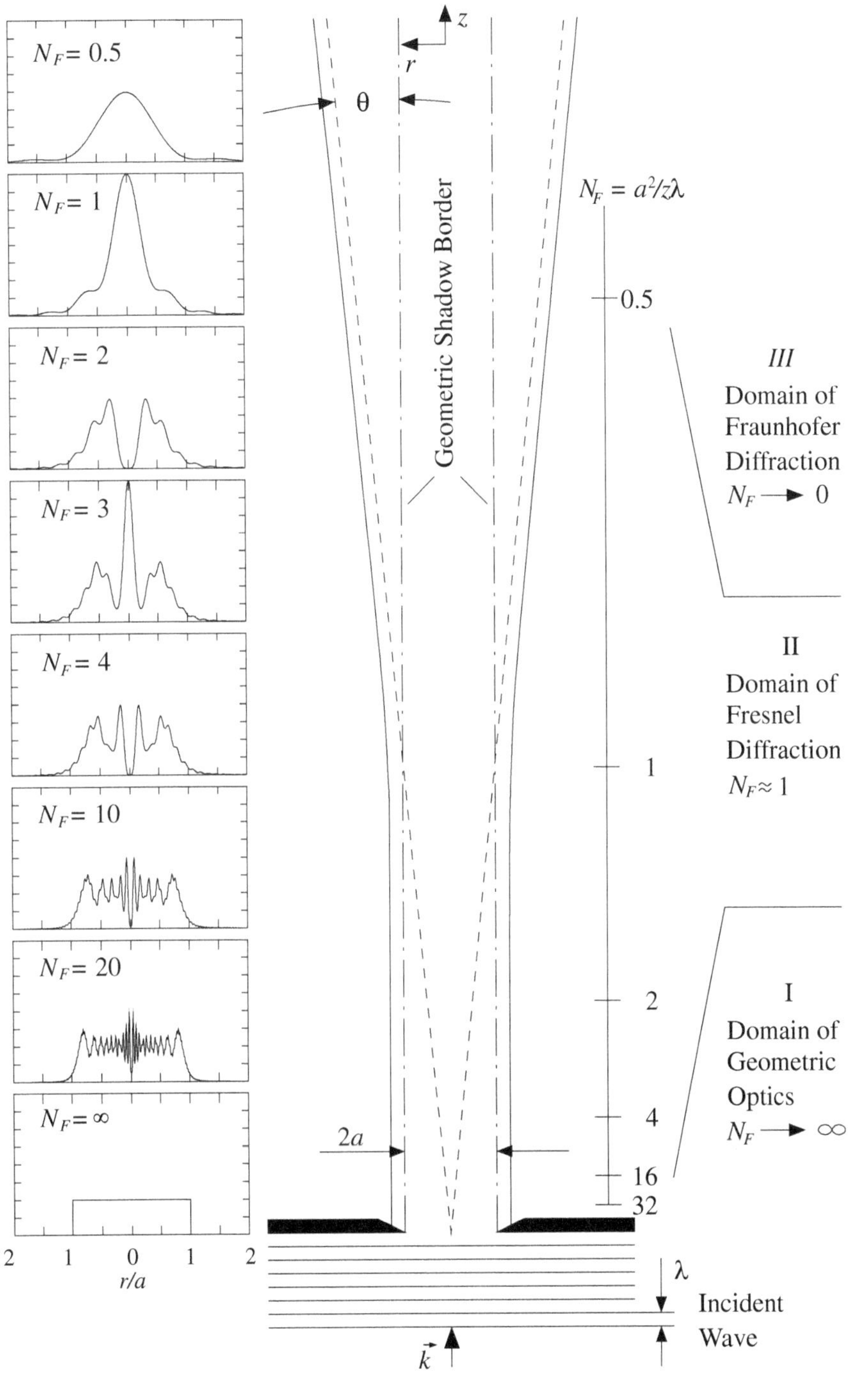

Fig. 4.2 Radiation propagation after diffraction at a circular aperture

depending upon distance. For this, the number of intensity maxima always corresponds to the Fresnel number.

In the following sections, the individual regions will be described in more detail.

4.2 Geometrical Optics

Most everyday optical experience falls within the area of geometrical optics: behind the illuminated object a shadow image originates, which corresponds to the geometrical projection of the object. This can be attributed to the dimensions of the observed objects being generally very large in relation to the wavelength of the visible light.[1]

Mathematically the transition to geometrical optics occurs by neglecting the finite sizes of the wavelengths, corresponding to the limit

$$\lambda \to 0, \quad k = \frac{2\pi}{\lambda} \to \infty. \tag{4.3}$$

4.2.1 *Fermat's Principle*

The wave equation in the form

$$\Delta \vec{E} - \frac{n^2}{c_0^2} \frac{\partial^2}{\partial t^2} \vec{E} = 0 \tag{4.4}$$

n Refractive index of the medium
c_0 Speed of light in vacuum

is chosen as an initial point to conduct the limit, whereas the speed of light in the medium is expressed by the refractive index and the speed of light in vacuum. In general, the refractive index is dependent upon the location, $n = n(x, y, z)$. However, the space dependence is normally weak, which means the refractive index barely changes over a wavelength and can be seen as constant. This limitation is also required so that the validity of the wave equation can be assumed in this form, since in the derivation of the wave equation, it was assumed that $\varepsilon = n^2 =$ const is valid (cf. Sect. 3.3.3).

---

[1]Examples: a window with an edge one-meter long, seen from a distance of five meters, $\lambda = 550$ nm: $N_F = 3.6 \times 10^5$. The sun, radius approx. 700,000 km, distance approx. 150 million km, $\lambda = 550$ nm: $N_F \cong 6 \times 10^{12}$.

As an approach to solve Eq. 4.4, the planar wave is initially used. The space dependence of the refractive index is now considered by introducing a space dependence of the amplitude as well as of the spatial phase term

$$\vec{E} = E_0(x, y, z)\vec{e}\mathrm{e}^{ikL(x,y,z)-i\omega t} \tag{4.5}$$

$\vec{e}$ Unit vector in direction of the electric field vector
L Coordinate in direction of propagation

It is assumed that neither polarization direction nor propagation direction are influenced such that k can be written as scalar, and the vector character of the electric field does not have to be considered any longer as well. It is essential that the space dependence of the phase is not introduced via a spatial variation of the wave number, but rather via a space-dependent compression or extension of the propagation coordinate L. The reason is that the limit for an infinitely large k should be carried out later.

When inserted in the wave equation (Eq. 4.5), the approach results in

$$\Delta E_0 + 2ik\vec{\nabla}E_0\vec{\nabla}L + ikE_0\Delta L - k^2E_0\left(\vec{\nabla}L\right)^2 + n^2k^2E_0 = 0. \tag{4.6}$$

After this equation is divided by k^2 and the limit $k \to \infty$ is carried out, only the last two terms remain on the left side, and the so-called *Eikonal Equation*

$$(\nabla L)^2 = n(x, y, z)^2 \tag{4.7}$$

follows. The solution $L(x, y, z)$ of this equation is called *Eikonal*. Aside from the constant factor k, the Eikonal agrees with the spatial phase of the wave so that L = const is valid for surfaces of constant phase. Since the vector ∇L is perpendicular on these surfaces and points in the propagation direction of the wave, it can be interpreted as a "light beam."

The Eikonal can be written in the form

$$L(P_1 \to P_2) = \int_{s_1}^{s_2} n[\vec{x}(s)]\mathrm{d}s, \vec{x}(s_1) = \overrightarrow{OP}_1, \vec{x}(s_2) = \overrightarrow{OP}_2. \tag{4.8}$$

P_1, P_2 Starting point and end point of a light path

Since there are any number of different paths from P_1 to P_2, the propagation of the light beam has not yet been determined. The unique solution is derived by using of *Fermat's Principle*[2]:

[2]According to P. Fermat (1650).

$$\delta L(P_1 \to P_2) = \delta\left(\int_C n[\vec{x}(s)]\mathrm{d}s\right) = 0, \quad C : P_1 \to P_2\,. \tag{4.9}$$

It says that the optical path for a light beam from point P_1 to point P_2 always takes on an extremum, which means a minimum or a maximum. The optical path is the integral of the refractive index along the geometrical path of the light beam. If the refractive index is constant in the observed region, then the light beam always chooses the geometrically shortest path from P_1 P_2. If the refractive index continuously varies, then the geometrical path of the light beam can be curved.

Equation 4.9 is a short form to write the well-known criterion that an extremum of a function is present where its first derivation disappears. In this case, the derivation occurs according to a parameter ε, which characterizes the deviation of any other path C' from P_1 to P_2 from the minimal, optical path C

$$\delta L = \left.\frac{\mathrm{d}F}{\mathrm{d}\varepsilon}\right|_{\varepsilon=0}, F(\varepsilon) = \int_C n[\vec{x}(s) + \varepsilon \cdot \delta\vec{x}(s)]\mathrm{d}s, \quad C : \vec{x}(s), C'(s) : \vec{x}(s) + \varepsilon \cdot \delta\vec{x}(s).$$

To derive Fermat's principle, the characteristic is needed for which a line integral always disappears from ∇L over any closed curve, according to Stoke's theorem

$$\oint_{\delta S} \vec{\nabla} L\, \mathrm{d}\vec{s} = \iint_S \underbrace{\vec{\nabla} \times \left(\vec{\nabla} L\right)}_{=0} \mathrm{d}\vec{A} = 0. \tag{4.10}$$

Fermat's principle can be extended in such a way that interfaces can also be considered upon which the refractive index makes a jump. In this case, it is required that the Eikonal $L(x, y, z)$ remains continuously on the interface; hence, the path of the light beam does not exhibit a jump there. The well-known laws for reflection and refraction can then be derived from this condition.

4.3 Reflection and Refraction

If the refractive index remains constant over macroscopic distances s_j, then the integral for the optical path can be replaced by a sum

$$s = \sum_j s_j, n = n_j = \text{const. auf} \to \text{on} s_j : \int_s n\, \mathrm{d}s \to \sum_j n_j s_j. \tag{4.11}$$

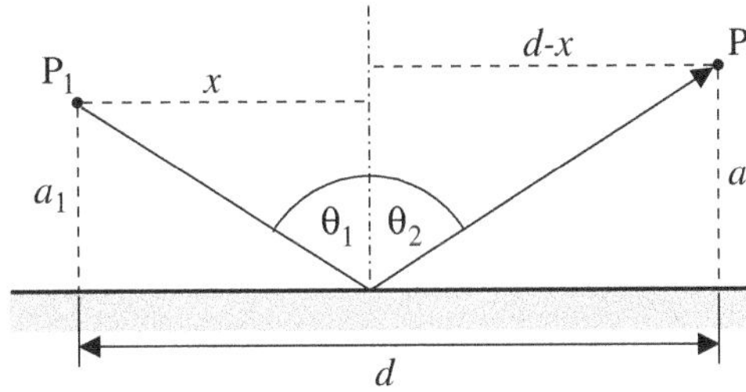

Fig. 4.3 Reflection at an interface

4.3.1 *Law of Reflection*

If a light ray propagates from P_1 to P_2 via reflection at an interface, the optical path is thus separated in two sections, one before and one after the reflection. The refractive index is the same on both sides of the path, since both lie in the same medium. Fermat's principle can then be read as follows:

$$\delta(ns_1 + ns_2) = 0, \tag{4.12}$$

whereas both parts of the path are given by

$$s_1 = \sqrt{a_1^2 + x^2} \text{ und } s_2 = \sqrt{a_2^2 + (d-x)^2}. \tag{4.13}$$

The optical path takes on an extreme value, if the condition

$$\frac{\mathrm{d}}{\mathrm{d}x}(ns_1 + ns_2) = n\frac{\mathrm{d}}{\mathrm{d}x}\left(\sqrt{a_1^2 + x^2} + \sqrt{a_2^2 + (d-x)^2}\right) = 0 \tag{4.14}$$

is fulfilled. From this follows

$$\frac{x}{s_1} = \frac{d-x}{s_2} \quad \Leftrightarrow \quad \sin\theta_1 = \sin\theta_2, \tag{4.15}$$

hence, the well-known law of reflection, (Fig. 4.3),

$$\theta_1 = \theta_2 \quad \text{Angle of incidence is equal to angle of reflection.} \tag{4.16}$$

4.3.2 *Law of Refraction*

The refraction law can be derived in a similar way. In this case, a light ray from P_1 in the medium with the refractive index n_1 should reach through the interface to P_2 in the medium with refractive index n_2 (Fig. 4.4).

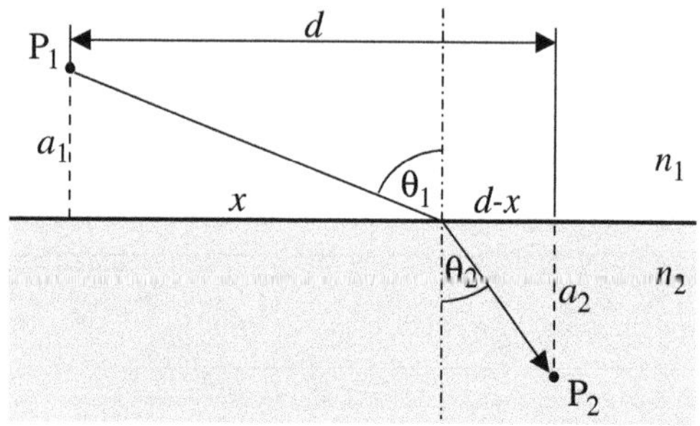

Fig. 4.4 Refraction at a plane interface

Fermat's principle now reads as follows

$$\delta(n_1 s_1 + n_2 s_2) = 0 \text{ with } s_1 = \sqrt{a_1^2 + x^2}, s_2 = \sqrt{a_2^2 + (d-x)^2}, \tag{4.17}$$

and by taking an analog derivation from the last section, Snell's law follows:

$$n_1 \sin\theta_1 = n_2 \sin\theta_2. \tag{4.18}$$

4.3.3 *Total Reflection*

An important limiting case of refraction can occur when a light ray falls upon the interface to an optically thinner medium, which means when $n_1 > n_2$. Since, due to Eq. 4.18, $\theta_2 > \theta_1$, the case occurs for a certain angle of incidence $\theta_1 = \theta_T$ that $\theta_2 = \pi/2$:

$$\theta_2 = \frac{\pi}{2} \Rightarrow \sin\theta_1 = \sin\theta_T = \frac{n_2}{n_1}. \tag{4.19}$$

For an angle of incidence larger than θ_T, there is no longer a solution for Eq. 4.18: no longer can radiation reach into the optically thinner medium. Therefore, the radiation is completely reflected. This case is called, hence, *total reflection*, θ_T is the *critical angle for total reflection* (Fig. 4.5).

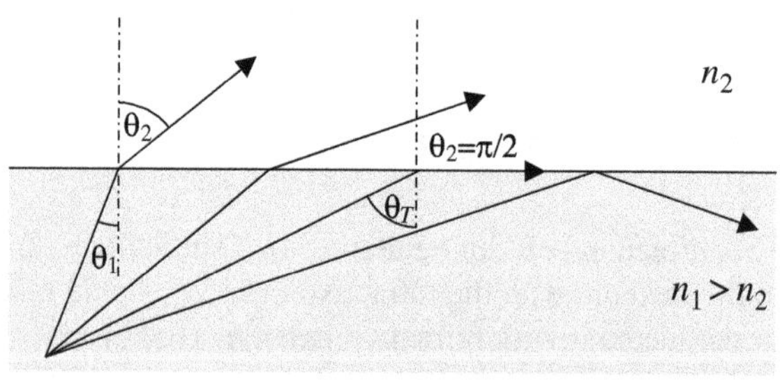

Fig. 4.5 Total reflection upon an interface to an optically thinner medium

The total reflection has many technical applications because it enables reflections very low in losses. The most important application is found in optical fibers. The optical core of the fiber consists of a material that is extremely transparent for the desired wavelength. It is surrounded by the fiber sheath which exhibits a significantly lower refractive index.[3] In this way the light can be transported through the fiber by repeated total reflection.

4.4 Transmission and Reflection Coefficients

The laws derived previously describe the direction of the radiation propagation when reflected and refracted on interfaces, but not the intensity relation of the reflected and refracted waves. As a rule, a light wave hitting an interface is neither completely reflected, nor completely transmitted under the angle of refraction. Both processes occur simultaneously, whereas the distribution of the intensity upon the reflected and transmitted wave is a function of three determining variables: that of the angle of incidence, of the refractive index behavior and of the polarization of the incident wave.

The ratio of the amplitudes is used, on the one hand, to characterize the reflection and transmission

$$r = \frac{E_r}{E_0}, t = \frac{E_t}{E_0}. \tag{4.20}$$

E_0 Electric field intensity of the incident wave
E_r Field intensity of the reflected wave
E_t Field intensity of the transmitted wave

r and t are the *amplitude reflection coefficient* and the *amplitude transmission coefficient*, respectively. On the other hand, the description over the intensity behavior is also common

$$R = \frac{I_r}{I_0}, T = \frac{I_t \cos\theta_t}{I_0 \cos\theta_0}, \tag{4.21}$$

I_0, I_r, I_t Intensities of the incident, reflected and transmitted waves
θ_0, θ_t Angle of incidence and angle of refraction

whereas R is the *reflectance* and T is the *transmittance*. The angles to the surface normals are contained in the relation for the transmittance, since the projection surface of the rays changes on the interface depending upon the angles.

[3] The step-index fiber is described here. There are also the so-called graded-index fibers, in which the refractive index declines continuously as it approaches the border. Then the light beam is not reflected sharply at the border, but rather is diverted by to the fiber center in a soft curve.

The values in Equations 4.20 and 4.21 as well as the most important relations between them are described in the following section. Absorption on the interface or in the medium is disregarded here. Equally, the magnetic characteristics of the materials are also not considered; thus, in the following, it is always assumed that for the magnetic susceptibility, $\mu = 1$ is valid in both media.

4.4.1 The Fresnel Equations

A complete description of the reflection and transmission coefficients, which is valid over a wide range of the most different materials, is known under the name the *Fresnel Equations*. For a plane wave hitting a smooth interface, these equations can be derived directly from the continuity conditions for the electric and magnetic fields on the interface. The continuity conditions result immediately from the Maxwell equations. The derivation in detail will not be covered here.

In particular, the Fresnel equations describe the polarization dependency of reflection and transmission. For this, the *s-polarization* and *p-polarization* are selected as independent polarization directions

$$\begin{array}{lll} \vec{E}\perp \text{ Plane of incidence} & \rightarrow & \text{s-Polarization,} \quad r_\perp, t_\perp \\ \vec{E}|| \text{ Plane of incidence} & \rightarrow & \text{p-Polarization,} \quad r_{||}, t_{||}. \end{array} \tag{4.22}$$

The indices $\perp$ and $||$ indicate the orientation of the E-field vector relative to the *plane of incidence* (Fig. 4.6). The plane of incidence is the plane that is spanned by the propagation direction of the incident and reflected wave; therefore, it always stands perpendicular on the interface. From both of these polarization directions, every other polarization can be constructed using superposition.

For the amplitude reflection and transmission coefficients in the s-polarization, the Fresnel equations result in

$$r_\perp \equiv \left(\frac{E_r}{E_0}\right)_\perp = \frac{n_1 \cos\theta_1 - n_2 \cos\theta_2}{n_1 \cos\theta_1 + n_2 \cos\theta_2}, \quad t_\perp \equiv \left(\frac{E_t}{E_0}\right)_\perp = \frac{2n_1 \cos\theta_1}{n_1 \cos\theta_1 + n_2 \cos\theta_2}. \tag{4.23}$$

For the p-polarization, they read as follows:

$$r_{||} \equiv \left(\frac{E_r}{E_0}\right)_{||} = \frac{n_2 \cos\theta_1 - n_1 \cos\theta_2}{n_1 \cos\theta_2 + n_2 \cos\theta_1}, \quad t_{||} \equiv \left(\frac{E_t}{E_0}\right)_{||} = \frac{2n_1 \cos\theta_1}{n_1 \cos\theta_2 + n_2 \cos\theta_1}. \tag{4.24}$$

The behavior of the reflection and transmission coefficients according to the Fresnel equations is rendered in Fig. 4.7 or Fig. 4.8, respectively.

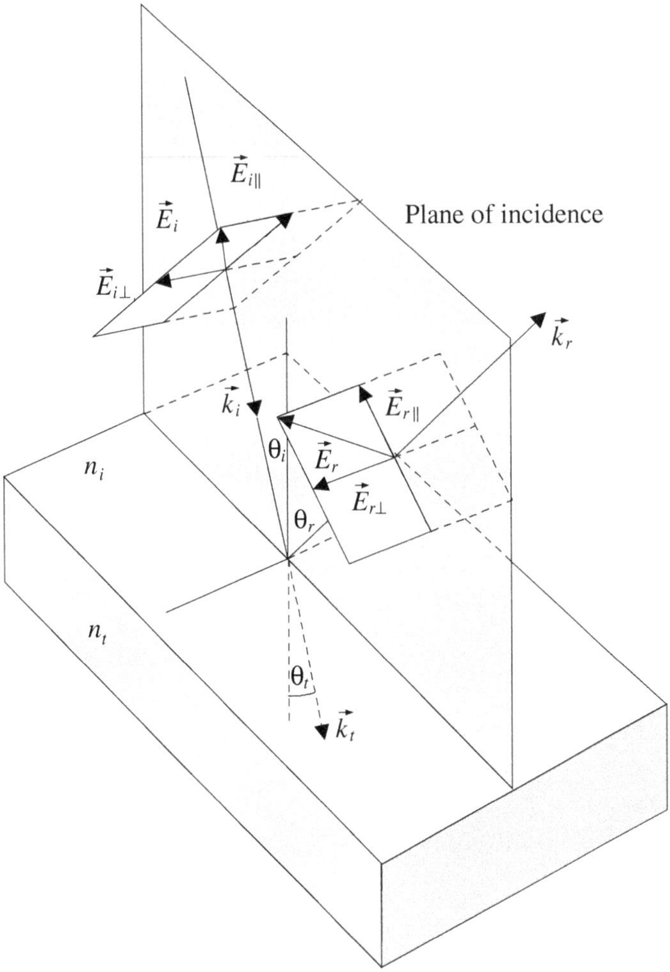

Fig. 4.6 On the geometry of a light wave hitting an interface. All three wave vectors lie in the plane of incidence. The E-field vectors stand perpendicular to the wave vectors and are decomposed into parts perpendicular or parallel to the plane of incidence

Important limits are the coefficients for *vertical incident light*:

$$\theta_1 = 0 \Rightarrow r_{\|} = -r_{\perp} = \frac{n_2 - n_1}{n_2 + n_1}, \quad t_{\|} = t_{\perp} = \frac{2n_1}{n_2 + n_1}. \tag{4.25}$$

The negative sign means that the light wave undergoes a phase jump of π there:

$$\vec{E}_0 \sim \cos\left(\vec{k}_0\vec{x} - \omega t\right) \rightarrow \vec{E}_r \sim \cos\left(\vec{k}_0\vec{x} - \omega t + \pi\right). \tag{4.26}$$

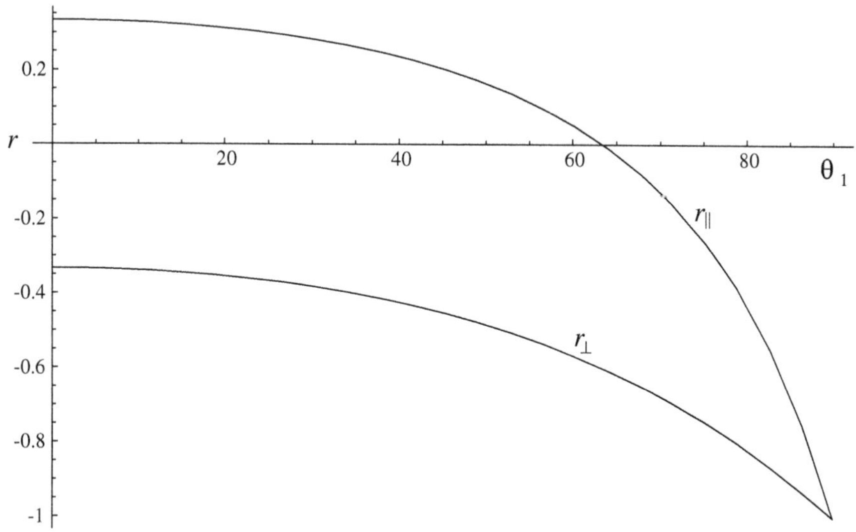

Fig. 4.7 Amplitude reflection coefficients according to Fresnel equations, for $n_1 = 1$ and $n_2 = 2$. The negative sign signifies a phase jump at the reflection by π

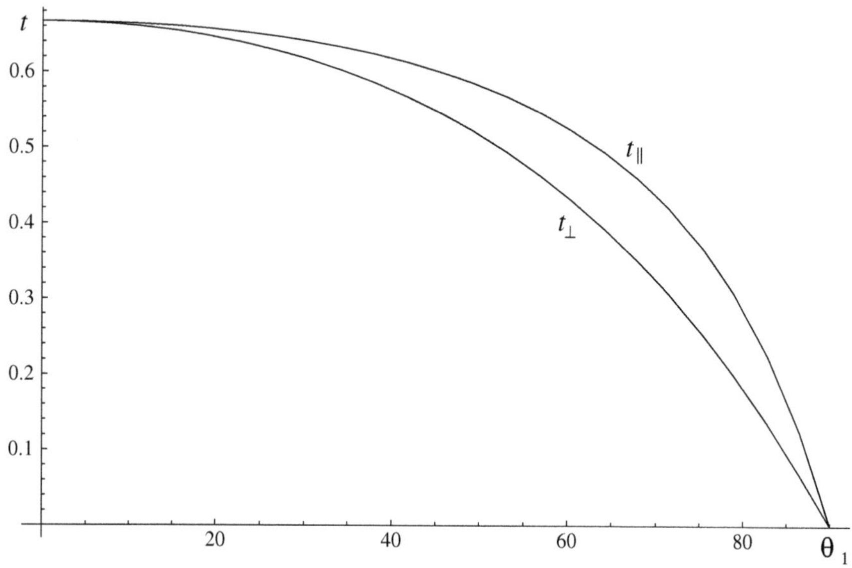

Fig. 4.8 Amplitude transmission coefficients ($n_1 = 1$, $n_2 = 2$)

In general, for the angle of incidence, the following holds

$$t_{\perp} - r_{\perp} = 1 \; \forall \theta_1, \tag{4.27}$$

whereas a similar relation for the p-polarization is valid, but only with vertical incidence

$$\theta_1 = 0 \Rightarrow t_{||} + r_{||} = 1. \tag{4.28}$$

4.4.2 *Reflectance and Transmittance*

For most practical applications, the intensity of the radiation is a more important reference parameter than the field intensity. It is then favorable to formulate the reflection and the transmission in relation to the intensities. This occurs by means of the *reflectance* and *transmittance* (Eq. 4.21).

The intensity describes the amount of energy transported by the light wave that penetrates the cross-sectional area A per unit of time. The cross-sectional area of the incident beam on the interface is given by the projection of A under the angle of incidence θ_1: $A\cos\theta_1$. The power hitting the interface is then the product of this area with the incident intensity; this is similar for the reflected and transmitted beam. Reflectance and transmittance are defined as the *ratio of the reflected or transmitted power, respectively, to the incident power*, such that

$$R \equiv \frac{I_r A\cos\theta_1}{I_0 A\cos\theta_1} = \frac{I_r}{I_0}, \quad T \equiv \frac{I_t A\cos\theta_2}{I_0 A\cos\theta_1} = \frac{I_t\cos\theta_2}{I_0\cos\theta_1} \tag{4.29}$$

follows. The cosine terms do not apply for the reflectance since the angle of reflection always corresponds to the angle of incidence.

Dependent upon the field intensity, the intensity is given by

$$I = \frac{\varepsilon_0 \varepsilon c}{2}|E|^2 = \frac{\varepsilon_0 c_0 n}{2}|E|^2 \text{ due to } c = \frac{c_0}{n}, \quad n = \sqrt{\varepsilon}. \tag{4.30}$$

In this way, R and T can now be expressed using the amplitude coefficients r and t

$$R_{\perp,||} = \frac{n_1|E_r|^2}{n_1|E_0|^2} = r^2_{\perp,||} \text{ und } T_{\perp,||} = \frac{n_2|E_t|^2\cos\theta_2}{n_1|E_0|^2\cos\theta_1} = \frac{n_2\cos\theta_2}{n_1\cos\theta_1}t^2_{\perp,||}. \tag{4.31}$$

In Fig. 4.9 the behavior of the reflectance is plotted over the angle of incidence. By means of the relations of Eq. 4.31, it can be shown that for each angle of incidence

$$R_{\perp} + T_{\perp} = 1, \quad R_{||} + T_{||} = 1 \tag{4.32}$$

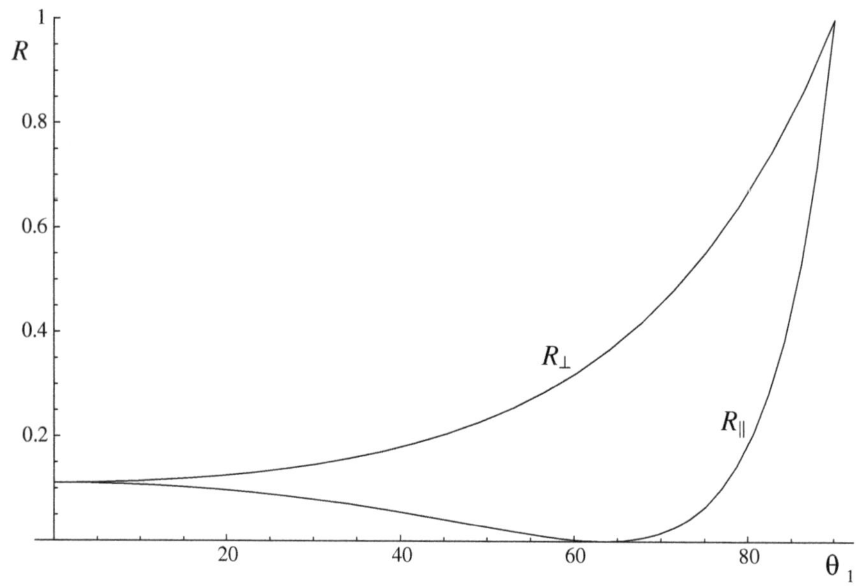

Fig. 4.9 The reflectance for s- and p-polarization over the angle of incidence ($n_1 = 1$, $n_2 = 2$)

is fulfilled. This is the expression for the energy conservation, since the irradiated energy has to disperse completely over the reflected and transmitted energy in the absence of absorption.

For perpendicular incidence, it follows that

$$\theta_1 = 0 \Rightarrow R_\perp = R_{||} = \left(\frac{n_2 - n_1}{n_2 + n_1}\right)^2, \quad T_\perp = T_{||} = \frac{4n_1 n_2}{(n_2 + n_1)^2}. \tag{4.33}$$

Therefore, light that falls perpendicularly on a glass plate is reflected by approximately 4 % ($n_1 = n_{\text{air}} = 1, n_2 = n_{\text{glass}} = 1.5$).

4.4.3 *The Brewster Angle*

Technically significant is the characteristic of the amplitude reflection coefficient to cut the zero line under a specific angle for the p-polarization (cf. Fig. 4.7): for the p-polarized light incident under this angle, there is no reflection, and the light is completely transmitted. This angle is called the *polarization angle* or the *Brewster angle*

$$\theta_1 = \theta_B \Rightarrow r_{||} = 0 \rightarrow R_{||} = 0, T_{||} = 1. \tag{4.34}$$

θ_B Brewster angle

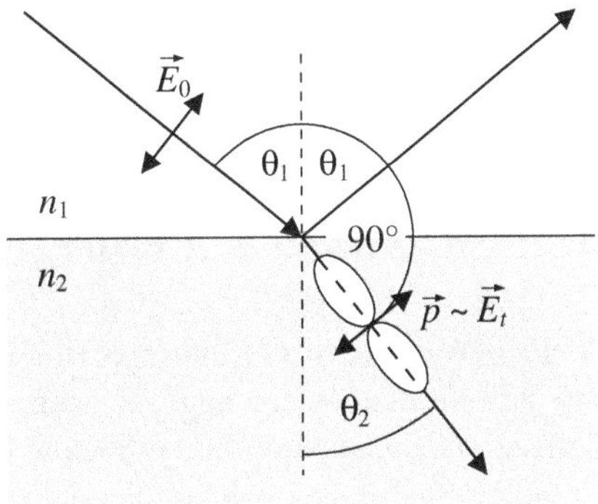

Fig. 4.10 Incident light and refraction under the Brewster angle. The dipole moments of the atoms in the material oscillate parallel to the electric field of the refracted wave. Since they only emit radiation perpendicular to the oscillation axis, the reflected wave disappears if it takes on an angle of 90° to the refracted wave

This characteristic is used, on the one hand, to polarize light and, on the other, to outcouple p-polarized light with low losses by using a glass plate arranged according to the Brewster angle, the so-called Brewster window.

The Brewster angle is located exactly where a refracted, transmitted beam and a reflected beam form a right angle, which means

$$\theta_1 = \theta_B : \theta_1 + \theta_2 = 90^\circ. \tag{4.35}$$

The disappearance of the reflection under this angle can be explained by the emission characteristic of excited dipoles (Fig. 4.10). The incident light wave forces the electrons of the atom in the lower material to oscillate around the atomic nucleus; the atoms can then be described by using oscillating dipoles. The oscillation direction corresponds to the polarization direction of the incident wave, which means it lies in the plane of incidence for p-polarized light. Like microscopic transmitting antennas, the oscillating dipoles emit radiation mainly perpendicular to their oscillating axis. Radiation emission does not occur parallel to the oscillation axis. For the p-polarized waves incident under the Brewster angle, the oscillation axis points exactly in the direction of the reflection. Since, emissions do not occur in this direction, reflected waves cannot originate.

4.5 Basic Optical Elements

Geometrical optics is based on the idea that light rays propagate in a straight line. On the interfaces boundary surfaces between the different materials, the law of reflection or the law of refraction applies. On this basis, the imaging behavior of the different optical elements, such as lenses, mirrors or prisms, can be described.

It is assumed that the ray propagates through air with the refractive index of $n_0 \cong 1$, and the optical elements exhibit a refractive index of $n > 1$.

4.5.1 Refraction at a Prism

A prism consists of two interfaces inclined toward each other (Fig. 4.11). A light ray is deflected by an angle δ as it passes through the prism. The amount of deflection depends upon the refractive index n and the vertex angle γ of the prism.

When Snell's law is used, the deflection angle results, in dependence upon the angle of incidence θ_1, in

$$\delta = \theta_1 - \gamma + \arcsin\left(\sin\gamma \cdot \sqrt{\frac{n^2}{n_0^2} - \sin^2\theta_1} - \cos\gamma \cdot \sin\theta_1\right). \tag{4.36}$$

The deflection is minimal when the ray passes through the prism symmetrically, which means that $\theta_1 = \theta_2 \equiv \theta_{sym}$. Then the following holds:

$$\begin{aligned} \theta_{sym} &= \arcsin\left(\frac{n\sin\gamma}{n_0\sqrt{1+2\cos\gamma}}\right) \text{ and } \delta = \gamma + 2\theta_{sym} - 180^\circ \\ &= 2\arcsin\left(\frac{n}{n_0}\sin\frac{\gamma}{2}\right) - \gamma. \end{aligned} \tag{4.37}$$

For a thin prism and nearly perpendicular incidence, Eq. 4.36 can be simplified to

$$\delta \cong \left(\frac{n}{n_0} - 1\right) \cdot \gamma. \tag{4.38}$$

In general, the refractive index n is a function of the wavelength. This characteristic is called *dispersion*. For *normal dispersion*, n declines with increasing

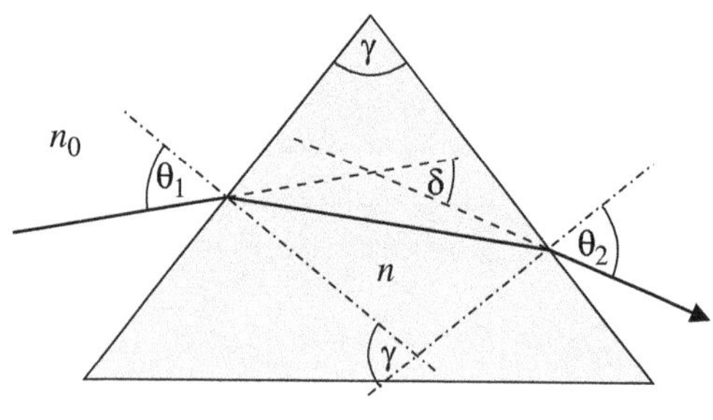

Fig. 4.11 Path of rays through a prism

wavelength.[4] A prism thus deflects short-wave radiation more strongly than long-wave radiation. Therefore, prisms can be used for spectral analysis of polychromatic radiation. For this, the *spectral resolution power* is

$$\frac{\lambda}{\Delta\lambda} = L_{\mathrm{eff}} \cdot \frac{\mathrm{d}n}{\mathrm{d}\lambda}. \tag{4.39}$$

$\frac{\lambda}{\Delta\lambda}$ Spectral resolution power
L_{eff} Effective (illuminated) base length of the prism
$\frac{\mathrm{d}n}{\mathrm{d}\lambda}$ Dispersion of the prism material

It indicates how large the distance $\Delta\lambda$ has to be between two spectral lines at the wavelength λ so that they can still be separated.

4.5.2 The Thin Lens

Lenses are transmitting optical elements defined by curved surfaces. As a rule, these interfaces are spherical surfaces.[5] Here a distinction is made between convex and concave surfaces: the concave surface is curved against, the concave surface is curved in the propagation direction of the light ray. By definition propagation occurs along the optical axis in a positive direction. Convex surfaces have thus a positive and concave surfaces a negative radius of curvature (Fig. 4.12).

According to the shape of both interfaces, six lens geometries can be distinguished: biconvex and biconcave lenses, plano-convex and plano-concave lenses, and positive and negative meniscus lenses (Fig. 4.13). There are, however, only two different functions that lenses can fulfill: they can function as *converging* or *diverging lenses*. Distinguishing between them is simple—the converging lenses are thicker in the middle than at the edges, whereas the diverging lenses are thinner at the center.[6]

When optical problems are treated theoretically, each lens will be characterized by its effect upon the light beam. The geometry of the lens is, therefore, not relevant; distinctions are only made between converging and diverging lenses. Practically, out of technical reasons, certain geometry can be preferred. The different geometries can also differ with respect to *imaging errors* (Sect. 4.7).

[4]There are also wave length regions in which *anomalous dispersion* occurs: here the refractive index increases with growing wavelength.

[5]As will be treated later in the section on image defects, the spherical surface only represents an approximation toward the real, ideal parabolic form of a lens. The reason for the spherical form lies in the significantly simpler manufacture.

[6]This rule is true as long as the refractive index of the lens material is larger than that of its surroundings.

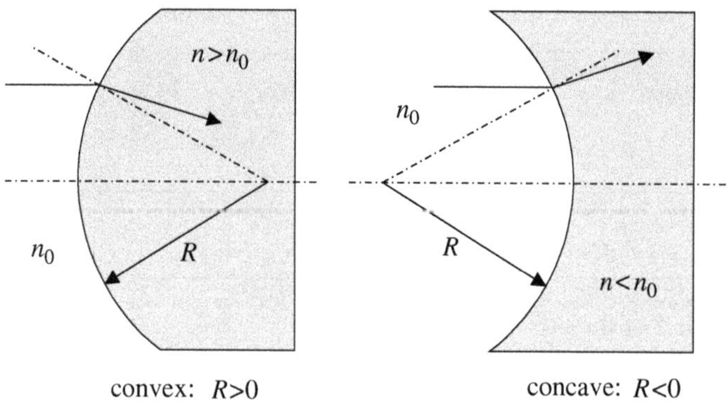

Fig. 4.12 For the definition of convex and concave surfaces

Fig. 4.13 The six different spherical lens geometries

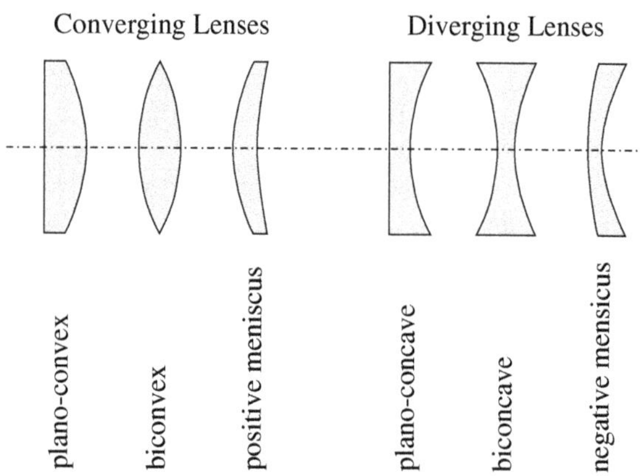

To begin with, the refraction of a thin lens is described in this section. Here, the designation of "thin" means that the optical path in the lens $n{\cdot}d$ can be neglected against the radius of curvature of the spherical surfaces. Geometrically, the lens can then be replaced by a plane in which the light ray is refracted. This plane is called the *principal plane* of the lens. For the construction of the principal plane of a lens see Fig. 4.15.

The thin lens can be replaced by a prism at the point where the light ray passes the lens, by forming respective tangents at the incident or exit plane. According to Eq. 4.38, the angle of refraction δ is then

$$\delta = \gamma \cdot (n - 1). \tag{4.40}$$

From Fig. 4.14 it can be seen that, in addition,

$$\delta = \phi_1 + \phi_2 \tag{4.41}$$

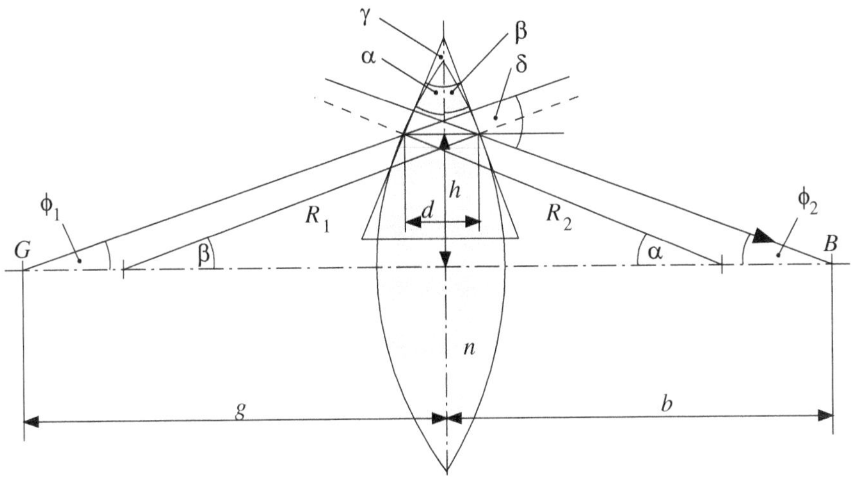

Fig. 4.14 The thin lens; construction of the angle of refraction

holds. For *paraxial rays*, ϕ_1, ϕ_2 and γ are small so that the approximations

$$\phi_1 = \frac{h}{g}, \phi_2 = \frac{h}{b} \text{ and } \gamma = \frac{h}{R_1} - \frac{h}{R_2} \tag{4.42}$$

g Object distance
b Image distance

are valid. Inserting those results in

$$\frac{1}{g} + \frac{1}{b} = (n-1)\left(\frac{1}{R_1} - \frac{1}{R_2}\right). \tag{4.43}$$

$R_{1,\,2}$ Radius of curvature of the lens surfaces
n Refractive index of the lens material

For incident, paraxial rays, this means $\phi_1 = 0$ or $g \to \infty$; the image distance is constant

$$g \to \infty : \frac{1}{b} \equiv \frac{1}{b_\infty} = (n-1)\left(\frac{1}{R_1} - \frac{1}{R_2}\right) =: \frac{1}{f}. \tag{4.44}$$

f is the *focal length* of the lens. The *focal point* lies in the distance f from the lens: there all paraxial incident rays cut the optical axis. Symmetrically, the second focal point lies in the distance $-f$ on the other side of the lens. *The fundamental lens equation for thin lenses* follows from Eq. 4.43

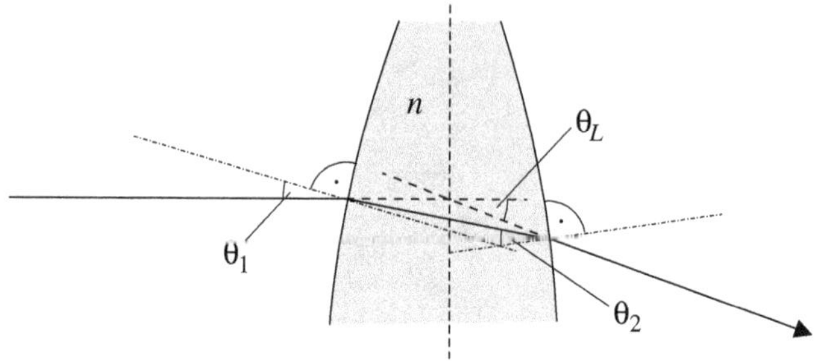

Fig. 4.15 On the construction of the position of the principal plane: it is located where the incident and the refracted ray intersect

$$\frac{1}{g}+\frac{1}{b}=\frac{1}{f}. \tag{4.45}$$

This equation describes the geometric-optical imaging behavior of every thin lens completely. Only the focal length f is required to characterize the lens.

From the definition in Eq. 4.44, it can be seen that the focal length can be positive as well as negative (Fig. 4.15)

$$\begin{array}{rl} R_1>0, R_2<0 \text{ or } R_1>0, R_2>0, R_2>R_1 & \\ \text{or } R_1<0, R_2<0, R_2<R_1 & \Rightarrow f>0 \\ R_1<0, R_2>0 \text{ or } R_1<0, R_2<0, R_2>R_1 & \\ \text{or } R_1>0, R_2>0, R_2<R_1 & \Rightarrow f<0 \end{array} \tag{4.46}$$

If $f > 0$, then the focal point lies *behind* the lens. There all paraxial incident rays intersect, which means that is where the light is focused. A converging lens is present. If, by contrast, $f < 0$, the focal point lies *in front of* the lens. In this point, the rays intersect for paraxial incidence, if they are extended backwards through the lens. Thus, paraxial incident rays diverge behind the lens, and this concerns a diverging lens.

When ray paths are constructed geometrically, the thin lens is represented by its *principal plane*. This plane is selected such that the refraction at both lens surfaces can be integrated into a change of direction at this plane alone. Furthermore, the focal length f of the lens has to be indicated.

Then, starting from an object point, three rays can be directly constructed

1. The *parallel ray* runs from the object point parallel to the optical axis. It is refracted at the principal plane such that it runs through the focal point f[7] (Fig. 4.16).
2. Starting from the object point, the *center point ray* intersects the optical axis exactly in the principal plane, which means it passes through the center point of

[7]In the case of a diverging lens, its elongation has to pass through the focal point, since it lies on the object side because $f < 0$.

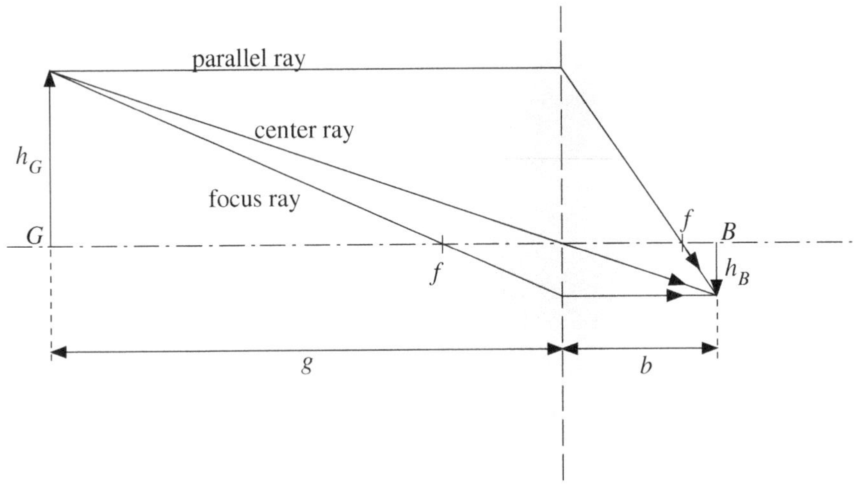

Fig. 4.16 Construction of the imaging of object G by a converging lens. The image B is real

the lens. This ray is not refracted at all: it runs exactly through the principal plane. This can be easily seen when one considers that every ray path is reversible.

3. The *focal ray* is exactly the reverse of the parallel ray. Starting from the object point, it intersects the optical axis at the second focal point $-f$. It is refracted at the principal plane such that it runs parallel to the optical axis behind the lens.

To construct the image point belonging to an object point, two of these rays are sufficient.

In the fundamental lens equation, Eq. 4.45, it can be seen that the image distance b can be negative according to the selection of the object distance g and the focal length f[8]

$$f > 0, g < f \text{ or } f < 0 \Rightarrow b < 0. \tag{4.47}$$

In this case, one speaks of a virtual image, but when b is positive, of a real image. For a real image, an image of the object originates at the position of the image, which can be recorded by illuminating a photographic plate, for example. For a virtual image, on the contrary, a photographic record cannot be made of the imaged object at the position of the image. When an additional lens is used, the virtual image can, however, be imaged such that a real image results. Hence, the virtual image is made visible for the human eye if one looks through the lens: the lens forms the virtual image in the eye as a real image on the retina. This principle is used, for example, to describe a magnifying glass.

---

[8]The object distance is always positive.

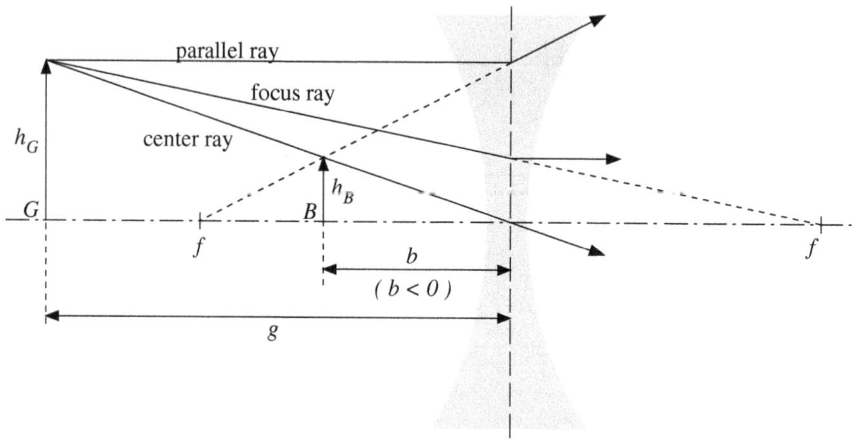

Fig. 4.17 Construction of the imaging of the object G by a diverging lens. In this case, the image B is virtual

The *lateral magnification* of the imaging is defined as the relation of the image size to the object size (Fig. 4.17):

$$V = \frac{h_B}{h_G} = -\frac{b}{g}. \tag{4.48}$$

h_G Object size
h_B Image size

When Eq. 4.45 is applied, the following results

$$V = \frac{-f}{g-f}, \tag{4.49}$$

which means that the lateral magnification depends only upon the object distance and the focal length. For $V > 0$ the image is upright; for $V < 0$, by contrast, the image is reversed.

4.5.3 *The Thick Lens*

For thick lenses the optical path of the light in the lens can no longer be neglected. Allowance is made for this by introducing two principal planes (Fig. 4.18).

The deflection of the light by the lens is described by the refraction at each rear principal plane. Focal length, object distance and image distance are measured from

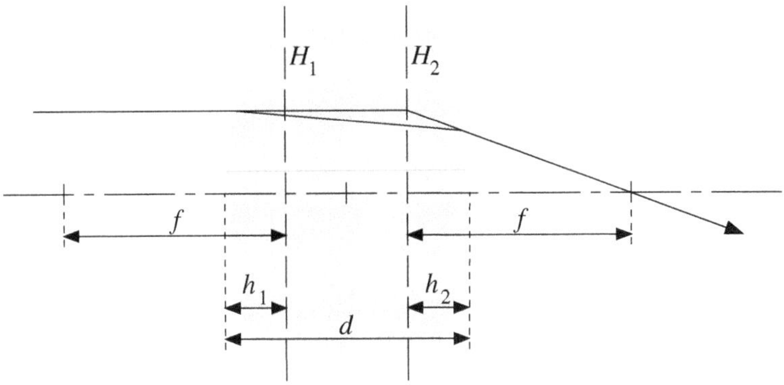

Fig. 4.18 The thick lens. The focal lengths are each measured from the closer principal plane. The refraction is carried out on the second principal plane H_2

the neighboring principal plane. With these assumptions, the optical equation, Eq. 4.45, still holds, whereas the focal length is now given by

$$\frac{1}{f} = (n-1) \cdot \left[\left(\frac{1}{R_1} - \frac{1}{R_2} \right) + \frac{n-1}{n} \frac{d}{R_1 R_2} \right]. \tag{4.50}$$

d Thickness of the lens

The first part of Eq. 4.50 exactly corresponds to the definition for the focal length of the thin lens. The second part factors in the thickness of the lens. The position of the principal planes is determined by their distance $h_{1,\,2}$ to the vertex of each of the lens surfaces

$$h_{1,2} = \frac{fd}{R_{1,2}} \frac{1-n}{n}. \tag{4.51}$$

To construct the image through a thick lens, the same ray trajectories can be used as for the thin lens. Between both principal planes, the rays are connected axially parallel to each other (Fig. 4.19).

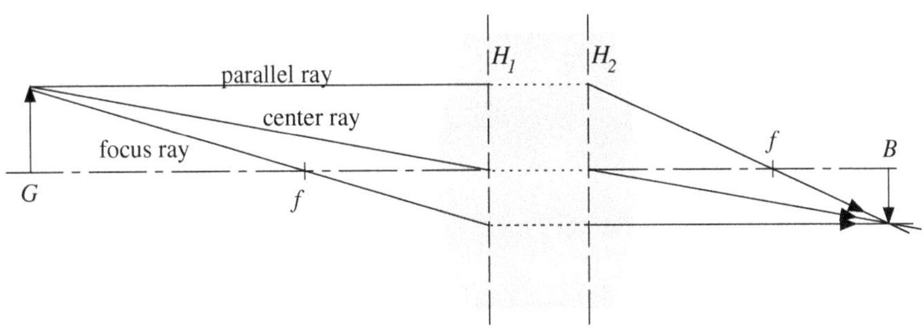

Fig. 4.19 Construction of the ray trajectory through a thick lens

4.5.4 Spherically Curved Mirrors

In geometrical optics, the propagation of light rays always occurs along the optical axis. Only the orientation of the optical axis is changed through a reflection on a planar mirror. Every optic system can be "folded" in such a way that the propagation occurs altogether along a fixed line. To accomplish this, the mirror simply has to be removed, or rather replaced by an aperture with the corresponding opening.

In addition to orienting the optical axis anew, curved mirrors generate an optical image. In order to obtain a continuous optical axis, the mirror has to be replaced by a transmitting optical element with the same imaging characteristics. In the case of a spherically curved mirror, this is a thin, spherical lens.

If the distance from the optical axis h is small in comparison to the radius of curvature R of the mirror, then the light ray hits the mirror surface at a small angle θ to the surface normal, and the approximation

$$\theta \cong \sin\theta = \frac{h}{R} \tag{4.52}$$

can be used. According to the law of reflection, Eq. 4.16, the ray is reflected under the same angle so that the reflected ray includes the angle $2\,\theta$ with the optical axis and cuts the optical axis at the distance (Fig. 4.20).

$$\Delta z = \frac{h}{2\theta} = \frac{R}{2} \equiv -f \tag{4.53}$$

This is, except for the sign, the focal length of the spherical mirror. For the sign of R, the previously known convention is used that the radius of curvature is positive for convex surfaces and negative for concave surfaces. Then, using the definition for the sign of the focal length conducted in Eq. 4.53, a spherical mirror with the radius of curvature R can be replaced directly by a thin lens with the focal length f.

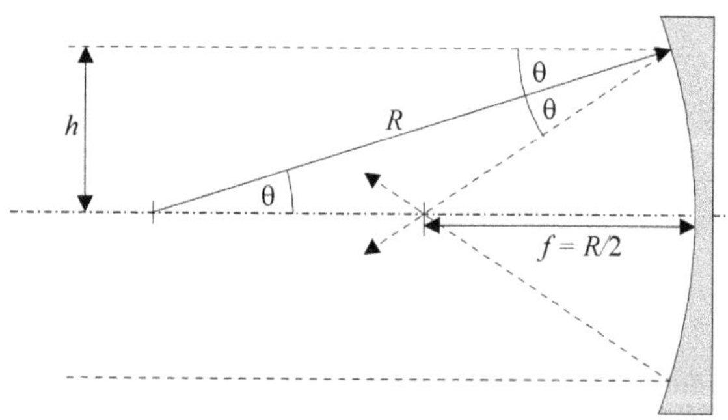

Fig. 4.20 Reflection on a spherical mirror

4.6 Matrix Formalism of Geometrical Optics

The propagation and imaging of light rays in geometrical optics can be represented in a compact form using matrices.

Under the assumption that all optical elements are rotational-symmetric regarding their optical axes, a light ray is completely described at every point by designating the distance h and the inclination θ to the optical axis. For the unaffected propagation by the route Δz from $P_1(h_1, \theta_1)$ to $P_2(h_2, \theta_2)$, the following is valid for small angles to the optical axis

$$\begin{aligned} h_2 &= h_1 + \Delta z \cdot \theta_1 \\ \theta_2 &= \theta_1 . \end{aligned} \tag{4.54}$$

If the $h_{1,2}$ and $\theta_{1,2}$ are combined into vectors, both of these equations can be represented as matrix equations

$$\begin{pmatrix} h_2 \\ \theta_2 \end{pmatrix} = \begin{bmatrix} 1 & \Delta z \\ 0 & 1 \end{bmatrix} \cdot \begin{pmatrix} h_1 \\ \theta_1 \end{pmatrix} . \tag{4.55}$$

This kind of *ray transfer matrices* can be specified for every image of geometrical optics

$$\vec{r}_2 = \mathbf{M} \cdot \vec{r}_1 \text{ with } \vec{r}_i = \begin{pmatrix} h_i \\ \theta_i \end{pmatrix}, \quad i = 1, 2. \tag{4.56}$$

The ray transfer matrices for the most important elements are listed in Table 4.1.

After the ray passes through several optical elements, the ray path is calculated by multiplying the ray transfer matrices of the individual elements in the series of the ray trajectory:

$$\vec{r}_n = \mathbf{M}_n \cdot \mathbf{M}_{n-1} \cdot \ldots \cdot \mathbf{M}_2 \cdot \mathbf{M}_1 \cdot \vec{r}_1 . \tag{4.57}$$

$\vec{r}_1$ Initial point
$\vec{r}_n$ End point

In this way, the imaging behavior can be determined, even that of complicated optical systems.

4.7 Aberration

In the previously described laws on the propagation of rays, the approximation for paraxial rays is assumed, which is justified for rays that

Table 4.1 The ray transfer matrices of the most important optic systems

	Optic system		Ray transfer matrix
1	Distance	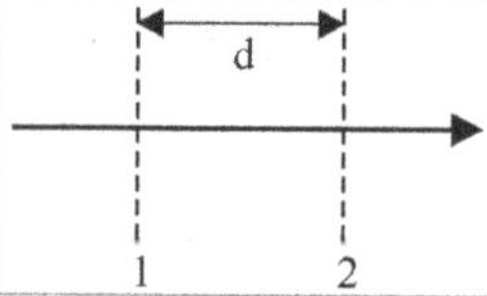	$\mathbf{M} = \begin{bmatrix} 1 & d \\ 0 & 1 \end{bmatrix}$
2	Refraction on plane surfaces	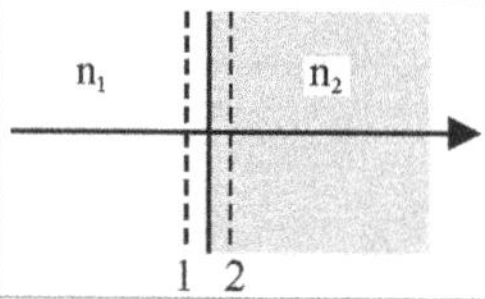	$\mathbf{M} = \begin{bmatrix} 1 & 0 \\ 0 & \frac{n_1}{n_2} \end{bmatrix}$
3	Refraction on spherical surfaces	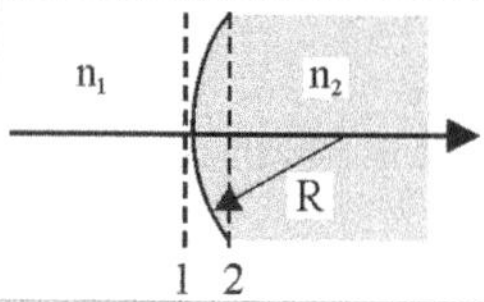	$\mathbf{M} = \begin{bmatrix} 1 & 0 \\ \frac{1}{R}\left(1 - \frac{n_1}{n_2}\right) & \frac{n_1}{n_2} \end{bmatrix}$
4	Lenses	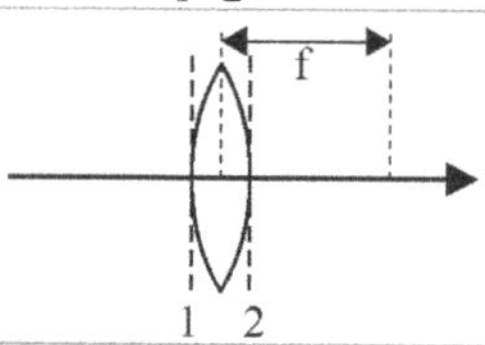	$\mathbf{M} = \begin{bmatrix} 1 & 0 \\ -\frac{1}{f} & 1 \end{bmatrix}$
5	Distance—lens—distance	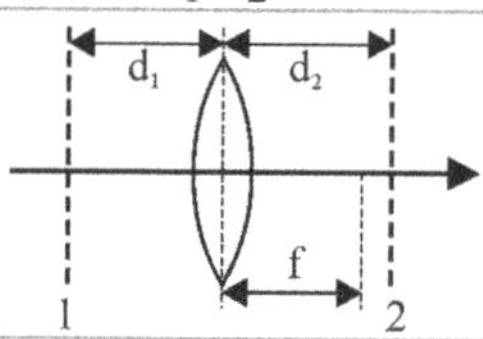	$\mathbf{M} = \begin{bmatrix} 1 - \frac{d_2}{f} & d_1 + d_2 - \frac{d_1 d_2}{f} \\ -\frac{1}{f} & 1 - \frac{d_1}{f} \end{bmatrix}$
6	Reflection on spherical mirror	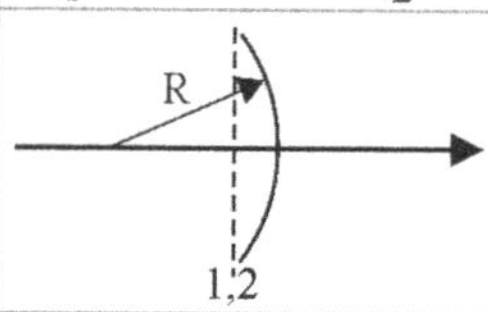	$\mathbf{M} = \begin{bmatrix} 1 & 0 \\ -\frac{2}{R} & 1 \end{bmatrix}$
7	Distance—spherical mirror—distance	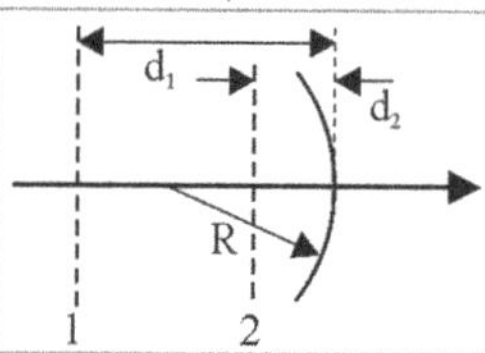	$\mathbf{M} = \begin{bmatrix} 1 - \frac{2d_2}{R} & d_1 + d_2 - \frac{2d_1 d_2}{R} \\ -\frac{2}{R} & 1 - \frac{2d_1}{R} \end{bmatrix}$
8	Distance—lens—distance—lens	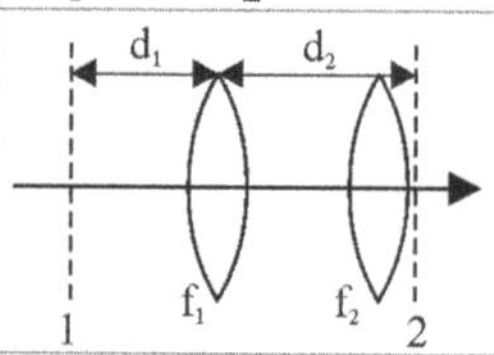	$\mathbf{M} = \begin{bmatrix} 1 - \frac{d_2}{f_1} & d_1 + d_2 - \frac{d_1 d_2}{f} \\ -\frac{1}{f_1} - \frac{1}{f_2} + \frac{d_2}{f_1 f_2} & 1 - \frac{d_1}{f_1} - \frac{d_1 + d_2}{f_2} + \frac{d_1 d_2}{f_1 f_2} \end{bmatrix}$

(continued)

Table 4.1 (continued)

	Optic system		Ray transfer matrix
9	Sequence of m identical lenses	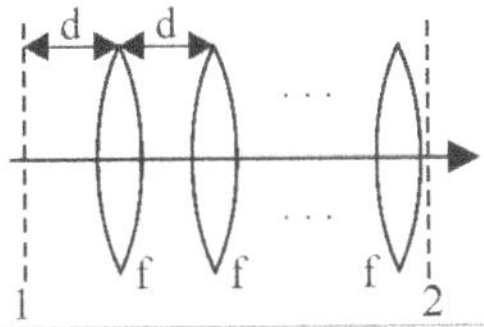	$\mathbf{M} = \begin{bmatrix} A & d \cdot B \\ -\frac{B}{f} & A - \frac{d}{f}B \end{bmatrix}$, $A = \sin m\varphi - \sin(m-1)\varphi$, $B = \sin m\varphi, \cos\varphi = 1 - d/(2f)$
10	Refractive medium with planar interfaces	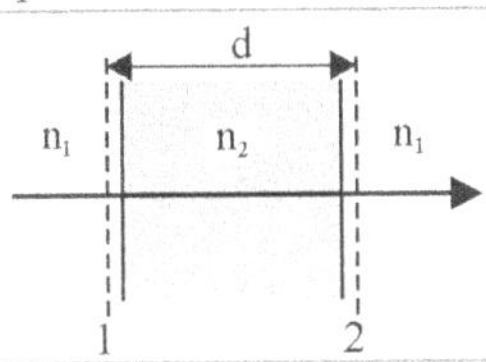	$\mathbf{M} = \begin{bmatrix} 1 & \frac{n_1 d}{n_2} \\ 0 & 1 \end{bmatrix}$
11	Thick lens	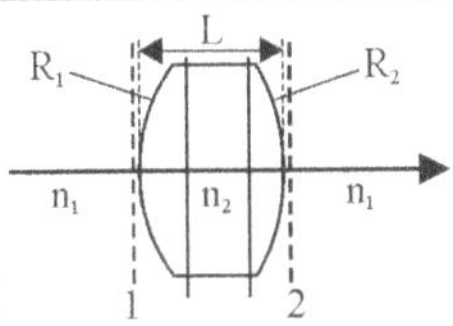 $R_1 > 0, \quad R_2 < 0,$ $n_2 > n_1$	$\mathbf{M} = \begin{bmatrix} 1 + \frac{h_2}{f} & L\frac{n_1}{n_2} \\ -\frac{1}{f} & 1 - \frac{h_1}{f} \end{bmatrix}$, $h_{1,2} = \frac{fL}{R_{2,1}} \frac{n_1 - n_2}{n_2}$, $\frac{1}{f} = \frac{n_2 - n_1}{n_1}\left(\frac{1}{R_1} - \frac{1}{R_2} + \frac{n_2 - n_1}{n_2}\frac{L}{R_1 R_2}\right)$
12	Thermal lens	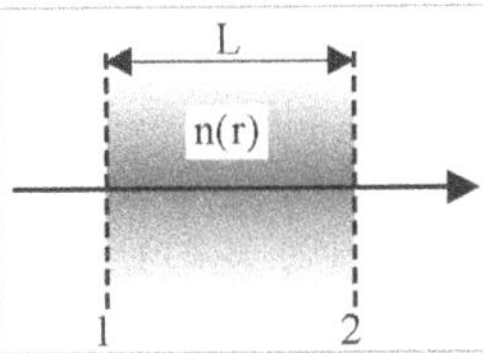	$n_2 = n_1(1 + \varepsilon r^2)$: $\mathbf{M} = \begin{bmatrix} 1 + \varepsilon L^2 & L/n_1 \\ 2\varepsilon L n_1 & 1 + \varepsilon L^2 \end{bmatrix}$

- only form a small angle to the optical axis,
- feature a small distance to the optical axis, and
- are monochromatic.

The worse these conditions are fulfilled, the more strongly aberrations occur. These aberrations are of principal nature; they are an expression for the deviation of the actual ray path from the geometric-optical model of paraxial rays. Technical deviations, or those caused by misalignment are not yet considered here.

Initially one distinguishes between *chromatic*[9] and *monochromatic* aberrations.[10] The chromatic aberrations result from the fact that the refractive index n is a function of the light frequency. The monochromatic aberrations also appear for single frequency light. Here, in turn, two groups differ: there are, on the one hand, aberrations that *blur* the image and, on the other, those that *distort* the image.

[9] Chromatic (Greek) colored; monochromatic: unicolored.

[10] Aberration (Latin): deviation, error.

For the most part, the paraxial description is based on the assumption that sin θ can be represented by θ with sufficient precision. At any rate, an improvement of the representation would result if from the series expansion of the sinus

$$\sin\theta = \theta - \frac{\theta^3}{3!} + \frac{\theta^5}{5!} - \frac{\theta^7}{7!} + \ldots \tag{4.58}$$

the following term of the third order were expanded to third order. The deviations that result from the theory of the first order are the five *monochromatic aberrations of the third order*

1. spherical aberration,
2. coma,
3. astigmatism,
4. curvature of the image field and
5. distortion.

The first complete, algebraic representation of these aberrations can be traced back to SEIDEL,[11] which is why these are called *Seidel's aberrations.*

The further terms in Eq. 4.58 lead to aberrations of a higher order, whose significance, however, declines more and more. The errors of the fifth order also play a role, when the deviations of the third order have been corrected by appropriate measures.

In general, aberrations can be understood in such a way that, for each ray originating from an object point, the position of the image point deviates from the ideal image point according to geometrical optics (Fig. 4.21). The lateral deviations Δx and Δy are expressed in dependence upon the position (r, φ) of the ray's intersecting point through the lens and upon the position (0, y_0) of the object point:

$$\begin{aligned}\Delta x &= Br^3 \sin\varphi - 2Fr^2 y_0 \sin\varphi\cos\varphi + Dry_0^2 \sin\varphi \\ \Delta y &= Br^3 \cos\varphi - 2Fr^2 y_0 (1+\cos^2\varphi) + (2C+D)ry_0^2\cos\varphi - Ey_0^3.\end{aligned} \tag{4.59}$$

(0, y_0)	Position of the object point
(r, φ)	Position of the intersecting point through the lens
(Δx, Δy)	Deviation of the real image point from the ideal image point
B, C, D, E, F	Seidel coefficients

The terms in Eq. 4.59 are ordered according to powers in r. The coordinate system has been selected in such a way that the optical axis runs in the z-direction and the object point lies on the y-axis. The Seidel coefficients are specific constants for each optical system; it is not necessary to know their numerical value to discuss the aberrations.

[11]Ludwig von Seidel, 1821–1896.

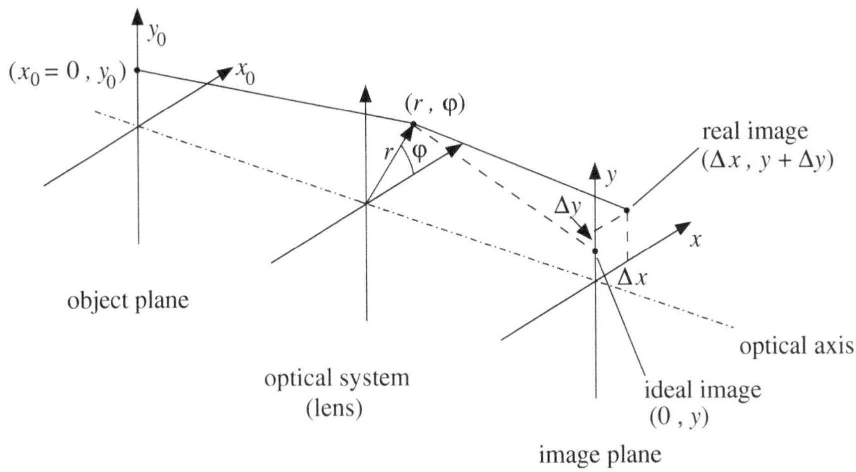

Fig. 4.21 Image of an object point behind an optical system in the image plane. The aberrations lead to the real image being shifted by (Δx, Δy) as compared to the ideal image according to geometrical optics

4.7.1 Spherical Aberration

In Sect. 4.5.2 it was already pointed out that the spherical form of common lenses only represents an approximation—selected for production related reasons—of the ideal, parabolic or hyperbolic form. Using the approximation $\sin\theta \approx \theta$, hence in the first order, both forms agree. Less paraxial rays hit the curved surface under larger angles, and the deviation from the ideal form becomes noticeable: rays that cross the lens further away from the optical axis do not intersect it at the focal point of paraxial rays. This deviation is called *spherical aberration*. It contributes the terms

$$\begin{aligned}\Delta x &= Br^3 \sin\varphi \\ \Delta y &= Br^3 \cos\varphi\end{aligned} \tag{4.60}$$

to the overall error. If the marginal rays intersect the optical axis in front of the focal point of paraxial rays, *positive spherical aberration* exists; as a rule this is seen in converging lenses. In another case, the marginal rays intersect the optical axis behind the focal point of paraxial rays. This *negative spherical aberration* is generally found in diverging lenses.

The spherical aberration leads to an object point not being imaged in focus on an image point, but rather on a disc; therefore, it counts among the aberrations that deteriorate the image. The diameter of the disc Δr_{SA} varies depending upon the distance to the lens

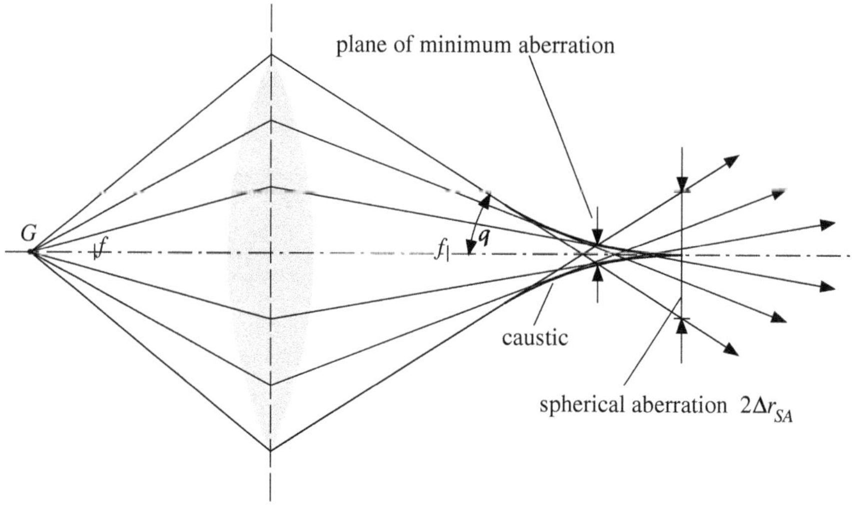

Fig. 4.22 Spherical aberration of a converging lens

$$\Delta r_{\mathrm{SA}} \sim \theta^3 \approx \left(\frac{r}{\Delta z}\right)^3. \tag{4.61}$$

Δz Distance to lens

The envelope of the refracted rays is called *caustic* (Fig. 4.22); through the intersecting point of the marginal rays and the caustic, the position is identified at which point the disc has the minimal diameter. This is the best position at which to observe the image.

Clearly, the spherical aberration can be reduced by narrowing the lens opening, since the error is proportional to r^3. From this, the amount of light passing through the lens is, however, reduced. Another method of reducing spherical aberration consists in selecting the appropriate combination of converging and diverging lenses.

4.7.2 Coma

The coma, or also asymmetry error, is an aberration that deteriorates the image and appears for object points that do not lie on the optical axis. The cause for the coma is that the principal "plane" of the lens can only be observed in the paraxial area as a plane, but is actually curved. This leads to different magnifications for rays with different distances r from the center of the lens. If the smallest magnification occurs

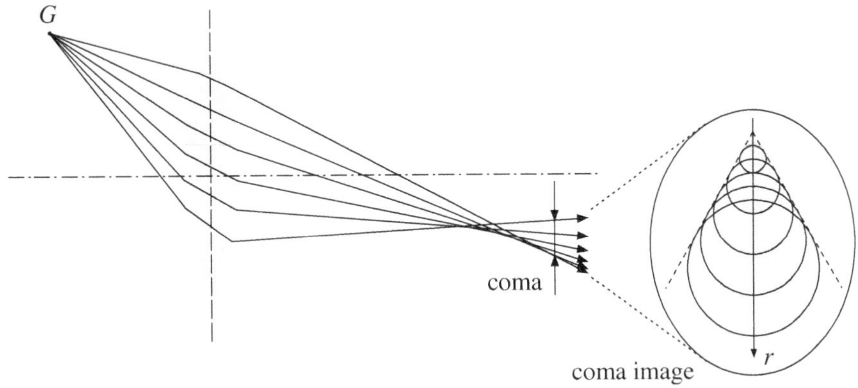

Fig. 4.23 Formation of the coma and the coma image. An object point is imaged on a cone, whose apex points to the optical axis at positive coma. h is the distance to optical axis. At negative coma, the cone lies with the apex outwards

for the marginal rays, the coma is thus negative; if, in contrast, the magnification is maximum, then the coma is positive.

The coma is described by the terms

$$\begin{aligned} \Delta x &= -2Fr^2 y_0 \sin\varphi\cos\varphi = -Fr^2 y_0 \sin 2\varphi \\ \Delta y &= -2Fr^2 y_0 (1+\cos^2\varphi) = -Fr^2 y_0 (2+\cos 2\varphi) \end{aligned} . \tag{4.62}$$

The cone-shaped image is typical for the coma, which produces a single object point. The rays that pass through the lens on a ring with the radius r produce a ring as an image, whose position and radius is proportional to r. The superposition of all rings results in a cone, whose apex is formed by the principal ray (Fig. 4.23).

4.7.3 *Astigmatism*

Astigmatism[12] is a further aberration that emerges for points that do not lie on the optical axis. Along with *field curvature* (see the next section), it contributes the following

$$\begin{aligned} \Delta x &= Dry_0^2 \sin\varphi \\ \Delta y &= (2C+D)ry_0^2 \cos\varphi \end{aligned} \tag{4.63}$$

[12]Astigmatism (Greek.): without a spot.

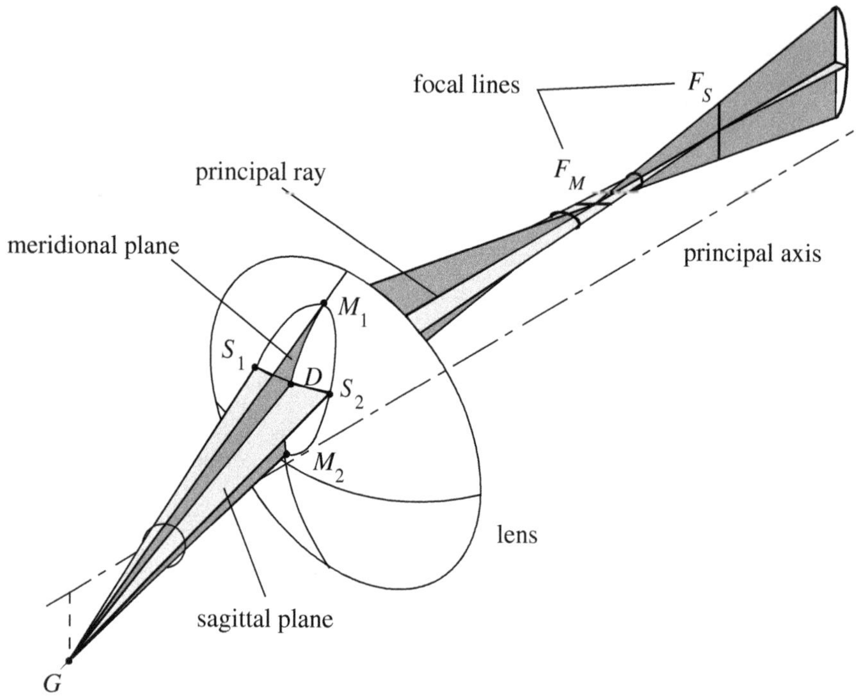

Fig. 4.24 Formation of astigmatism. The sagittal plane is represented in a light tone, the meridional plane in dark. D is the intersecting point of the principal ray through the lens

to the overall error. Astigmatism is caused when rays—leaving from the same object point and crossing the lens at the same distance r to the optical axis—hit the lens under different angles and are, therefore, refracted differently.

To simplify the description, two planes perpendicular to each other, along the propagation direction are defined. The *meridional plane* is spanned by the optical axis and the principal ray leaving the object point. The *sagittal plane* is then defined as the plane containing the principal ray and, in addition, standing perpendicular to the meridional plane.

Rays within the sagittal plane cross the lens symmetrically and are focused on a focal point, apart from the spherical aberration. The rays in the meridional plane hit the lens asymmetrically and are, therefore, focused on another point. In the case of a converging lens, the angle of incidence is enlarged for the meridional rays. As with the marginal rays for the spherical aberration, they are, therefore, focused on a point lying closer to the lens. For a diverging lens, the opposite effect appears, and the focus for the meridional rays lies further away from the lens. The object point is imaged on an ellipse, which contracts to a line at each of the focal points. In the focus of the meridional rays, this lens is perpendicular to the meridional plane; in the focus of the sagittal rays, it is perpendicular to the sagittal plane. In a plane that lies between both focal points, the object point is imaged on a circle (Fig. 4.24).

4.7.4 *Field Curvature*

Field curvature is closely related to astigmatism. Here, too, the cause is that the distance of the focusing point from the lens is dependent upon the inclination at which the bundle of rays hits the lens. Through this, object points—at the same distance to the lens, but with different distances to the optical axis—are imaged sharply on a curved image surface. If, in contrast, the image is projected upon a planar screen, the border area and central area are never sharp simultaneously.

The curvature of the image field is called the *Petzval field curvature* after the Hungarian mathematician JOSEF MAX PETZVAL. For converging lenses, the border areas of the image are curved inwards toward the lens, for diverging lenses from the lens outwards. The field curvature can be corrected by using an opposingly curved *field curvature flattening lens.*

4.7.5 *Distortion*

This last aberration of the third order is caused by the dependence of lateral magnification V upon the distance of the object point to the optical axis. This different magnification expresses itself in a distortion of the image in total (Fig. 4.25). The deviations belonging to this are described by

$$\begin{aligned} \Delta x &= 0 \\ \Delta y &= -Ey_0^3 \end{aligned} \quad . \tag{4.64}$$

For *positive* or *pincushion distortion*, the magnification for points lying far outwards is maximum, which means the distance from the optical axis is increased. The other way around, for *negative* or *barrel* distortion, the distance to the optical axis is reduced not to scale; the magnification is maximal for points lying very close to the optical axis.

As is case of coma and astigmatism, distortion can be influenced by positioning an aperture in the path of the ray. An aperture reduces the coma significantly; the influence on the distortion depends upon the position of the aperture: the further away the aperture is placed from the lens, the more pronounced the distortion is (Fig. 4.26).

The distortion can be balanced out by using a symmetric arrangement of two lenses with the aperture in the center. This way, the distortions cancel each other out through the effect of the individual lenses.

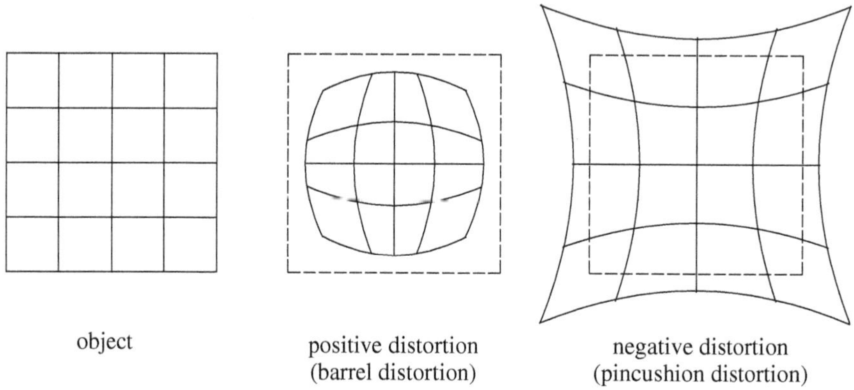

Fig. 4.25 Distortion of a square grid

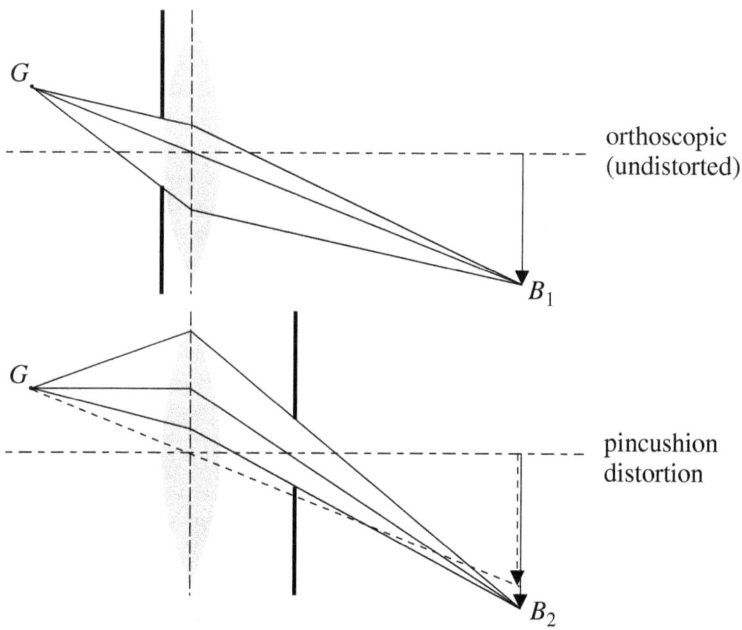

Fig. 4.26 The effect of aperture position on the distortion

4.7.6 *Chromatic Aberration*

Until now, only aberrations have been treated which mirror the deviation from the theory of paraxial rays and already appear for monochromatic light. In addition,

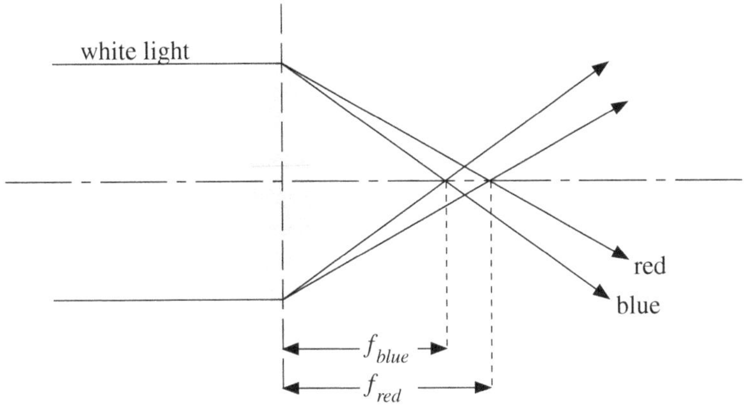

Fig. 4.27 Chromatic aberration. The lens has different focal lengths for different wavelengths

there are *chromatic aberrations* for polychromatic light.[13] These can even be significantly more serious than the monochromatic aberrations listed above.

Chromatic errors are caused by *dispersion*, which means the wavelength dependency of the refractive index. According to Eq. 4.44 the refractive index enters the focal length of the lens linearly. Thus, if the refractive index changes, then the lens exhibits a different focal length for different wavelengths; exactly as already represented for the prism in Sect. 4.5.1, a polychromatic light ray is split into its color components (Fig. 4.27). When a point is imaged by a lens, colored, concentric rings are produced. When measured over a specific wavelength interval, the radius of these rings is called a *lateral color aberration*.

Chromatic aberration can be compensated for by combining lenses appropriately. For this, two lenses are commonly used, made of different material, thus with different dispersion properties. Through the suitable selection of the materials and the focal lengths, a very good compensation of the chromatic aberrations can be attained. The corresponding corrective lens system is called *achromat*.

4.7.7 Diffraction Limit

As a conclusion, an aberration is discussed whose cause is not based on the approximation of paraxial rays, but rather on the most fundamental approximation for geometrical optics: the omission of the wavelength. Already, Sects. 4.1 and 4.2 pointed to the limitations of geometrical optics. This approximation is only valid if the dimensions of the imaged object are significantly larger than the wavelengths.

[13]polychromatic (Greek): multicolored.

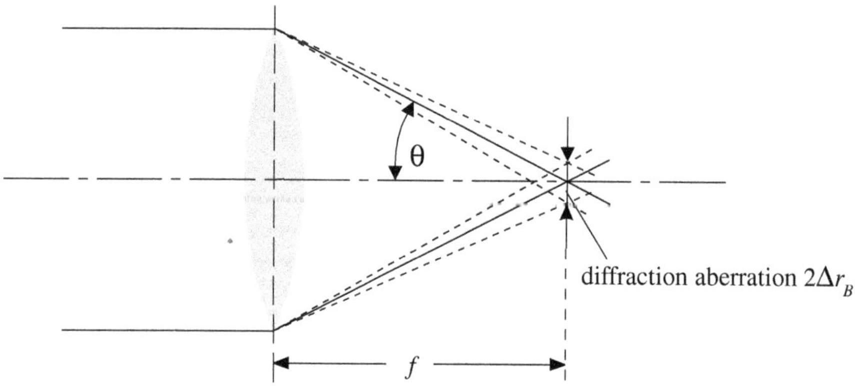

Fig. 4.28 Diffraction limit. Diffraction leads to a blurred image, the smaller the aperture angle θ is

The smaller the objects become, the stronger the wavelength of the light becomes noticeable, and *diffraction patterns* appear.

In the area of validity of geometrical optics, the diffraction effects are of secondary importance. Therefore, they can be considered as a form of a further aberration. *Aberration due to diffraction* is inversely proportional to the angular aperture of the optic, and leads to a blurring of the image when the angular aperture is *small*. The radius of the *diffraction pattern* which is also called *Airy disk*, upon which the aperture is imaged, is given by (Fig. 4.28)

$$\Delta r_B = \frac{\lambda}{2\theta}. \tag{4.65}$$

Δr_B Radius of the Airy disk
θ Angular aperture
λ Wavelength

The aberration due to diffraction behaves in exact opposition to that of Seidel's spherical aberration: the latter grows when the aperture angle becomes *larger*. Therefore, there is an *optimal aperture angle* θ_{opt}, for which the sum of the aberrations is minimal (Fig. 4.29). The diffraction and thus the mechanism of this error is treated in the following section.

4.8 Diffraction

If the Fresnel number is no longer very large, then the fundamental approximation of geometrical optics, $\lambda \rightarrow 0$, can no longer be justified. This is the case when the distance from the imaged opening is very large or the opening is no longer much larger than the wavelength. This is the region of *wave optics*.

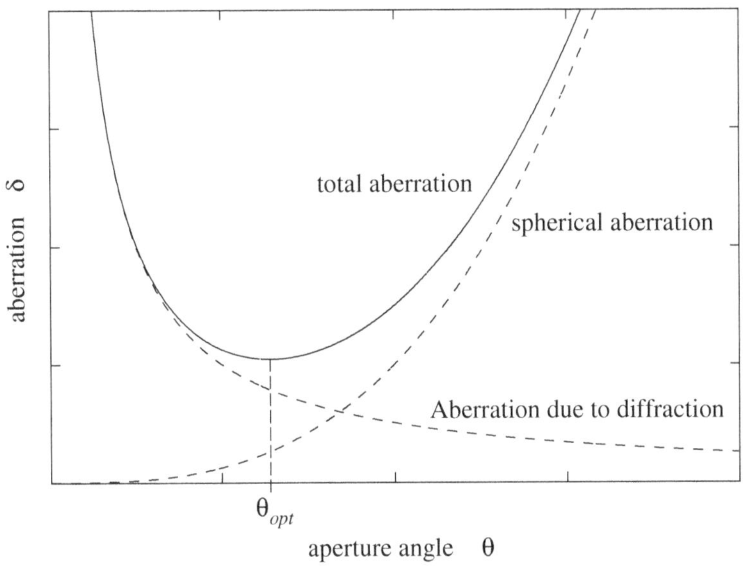

Fig. 4.29 Qualitative progression of the aberration as a function of the aperture angle. At θ_{opt} the total aberration made up of the diffraction and spherical aberrations is minimal

In wave optics, constructing geometric beam paths is no longer possible since the "light beam" no longer has any significance as a geometric line. Once again, the determination of the electromagnetic field utilizing the wave equation has to supersede the geometric construction.

Section 4.1 has already represented the foundational behavior of the beam beyond the geometric-optical region. The radiation field now extends over the geometric shadow boundary onto the shadow region, and the intensity distribution fluctuates in the beam interior. *Diffraction* causes this behavior.

4.8.1 Huygens' Principle and Kirchhoff's Diffraction Integral

An elementary approach to explain wave propagation can be traced back to the Dutch physicist Christiaan Huygens from the year 1690. Huygens' principle states that every point on a primary wave front is the initial point of secondary spherical waves, such that the primary wave-front is the envelope of the secondary waves at a later point of time. The secondary waves propagate at the same velocity and frequency as the primary wavefront (Fig. 4.30).

Huygens' principle was later rendered more precise by Fresnel. In Huygens' formulation, only those parts of the secondary waves are used that overlap with the envelope, whereas the rest of the spherical waves are ignored. Fresnel removed this

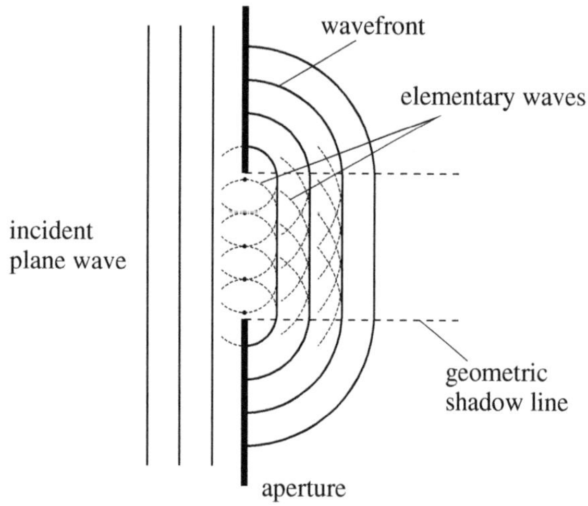

Fig. 4.30 The Huygens–Fresnel Principle. The wave entering from the *left* is diffracted at the aperture. Every point of the aperture opening can be seen as the source of spherical elementary waves, whose superposition results in the wave front behind aperture

inconsistency by introducing the interference principle. The Huygens–Fresnel principle, therefore, says that

> At every point of time all non-masked points of a wavefront are to be seen as sources of spherical secondary waves. The secondary waves propagate at the same velocity and frequency as does the primary wavefront. At every subsequent point the amplitude of the electromagnetic field is given by the superposition of these secondary waves. For the superposition, amplitude and phase of the waves have to be considered.

This principle serves as a basis to explain the diffraction phenomena. It also clarifies that it is principally impossible to separate the concepts interference and diffraction from one another.

A mathematical formulation of the Huygens–Fresnel principle is *Kirchhoff's diffraction integral*. For the field behind an aperture opening of the area A, it reads as follows[14]

$$E(\vec{r}) = \frac{ik}{2\pi} \iint_A E(\vec{r}') \frac{\exp(i\vec{k}\vec{R})}{R} \mathrm{d}A. \tag{4.66}$$

$E(\vec{r})$ Field intensity at the observation point behind the aperture
$E(\vec{r}')$ Field intensity in the aperture opening
$\vec{R} = \vec{r} - \vec{r}'$ Distance from the observation point to the aperture

[14]Indicated here is an approximation of Kirchhoff's diffraction integral for directed propagation. In a general case, the *inclination factor* $K(\theta) = \frac{1}{2}(1 + \cos\theta)$ has to be inserted into the integral, the factor allowing for the inclination towards the propagation direction of the primary wavefront. The inclination towards ensures that a wavefront does not propagate backwards but forwards, a point that has not yet been incorporated in the Huygen–Fresnel principle.

Fig. 4.31 An explanation of Kirchoff's diffraction integral

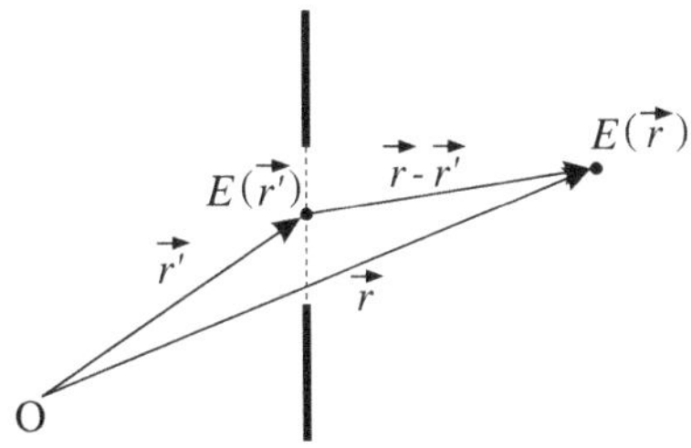

The resulting field intensity $E(\vec{r})$ at the observation point results from the superposition of the spherical waves starting from every point $\vec{r}'$ of the area A (Fig. 4.31).

Kirchhoff's diffraction integral can be derived from the scalar Helmholz equation, under the assumption that the field intensity and its gradient in the aperture opening correspond to the undisturbed field intensity of the incident wave, and the field outside the aperture opening disappears in the infinite. These assumptions are fulfilled for openings that are large compared to the wavelengths.

4.8.2 The Fresnel Diffraction

If the propagation direction of the wave only slightly deviates from the optical axis and if the distance to the aperture is large as compared to the aperture opening

$$R \gg a_x, a_y, \tag{4.67}$$

a_x, a_y Aperture radius in x or y direction

then the integrand in Eq. 4.66 can be simplified by using an approximation expression for R

$$R \approx |z - z'| + \frac{(x-x')^2 + (y-y')^2}{2|z-z'|}, \quad \vec{r} = \begin{pmatrix} x \\ y \\ z \end{pmatrix}, \vec{r}' = \begin{pmatrix} x' \\ y' \\ z' \end{pmatrix}. \tag{4.68}$$

This special paraxial approximation is called the Fresnel approximation. When this approximation is used, the *Fresnel diffraction integral* results from Eq. 4.66

$$E(x, y; R_0) = \frac{ik}{2\pi}\frac{e^{ikR_0}}{R_0} \iint_A E(x', y')\, e^{\frac{ik}{2R_0}\left\{(x-x')^2 + (y-y')^2\right\}} dx'dy'. \tag{4.69}$$

$R_0 = |z - z'|$ Paraxial distance to the aperture plane

$E(x, y; R_0)$ is the electric field in a plane with the distance R_0 to the aperture measured parallel to the optical axis. Often dimensionless position coordinates are introduced by normalizing upon the aperture dimensions a_x and a_y,

$$\xi = \frac{x}{a_x}, \eta = \frac{y}{a_y} \text{ and } \xi' = \frac{x'}{a_x}, \eta' = \frac{y'}{a_y}, \tag{4.70}$$

and separate Fresnel numbers are defined for the x- and y-direction

$$N_{F,x} = \frac{a_x^2}{\lambda R_0} \text{ und } N_{F,y} = \frac{a_y^2}{\lambda R_0}, \tag{4.71}$$

so that Eq. 4.69 can now be notated in this form

$$\begin{aligned} E(\xi,\eta) &= i\sqrt{N_{F,x}N_{F,y}}\,\mathrm{e}^{ikR_0}\int_{-1}^{1}\int_{-1}^{1} E(\xi',\eta')\mathrm{e}^{i\Phi(\xi,\xi',\eta,\eta')}\mathrm{d}\xi'\mathrm{d}\eta' \\ \text{mit } \Phi(\xi,\xi',\eta,\eta') &= \pi\left\{N_{F,x}(\xi-\xi')^2 + N_{F,y}(\eta-\eta')^2\right\}. \end{aligned} \tag{4.72}$$

The Fresnel approximation is normally used when $N_{F,\,x} \approx N_{F,\,y} \approx 1$ is valid for the Fresnel numbers, which means for a distance $R_0 \approx a_{x,y}^2 / \lambda$ from the aperture. The length of optical resonators mostly lie in this region.

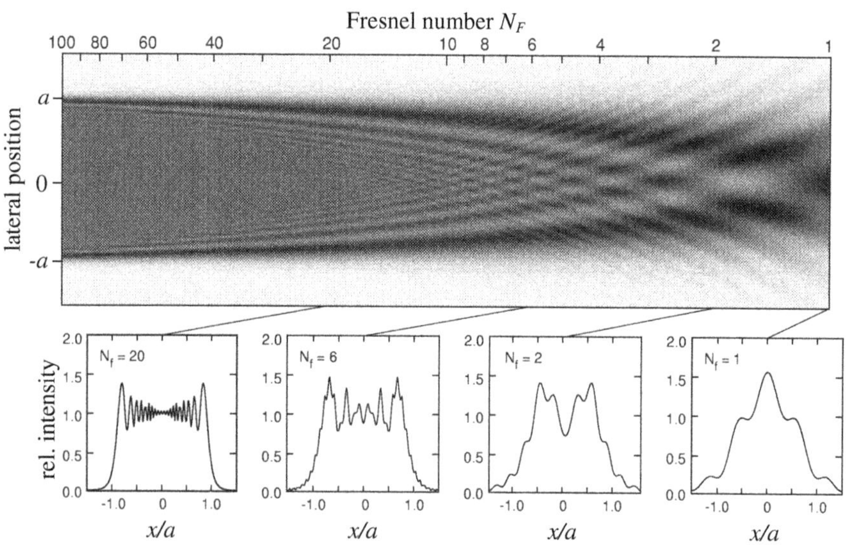

Fig. 4.32 Diffraction at a slotted aperture in the Fresnel region

Figure 4.32 shows the diffraction pattern behind the aperture for various Fresnel numbers. The number of diffraction maxima corresponds to the Fresnel number. For odd Fresnel numbers, the field distribution has, therefore, a maximum on the optical axis; for even Fresnel numbers, in contrast, a minimum. If one walks along the optical axis, the Fresnel number varies, resulting in an alternation of minima and maxima.

4.8.3 The Fraunhofer Diffraction

For very large distances to the aperture or very small Fresnel numbers, the phase expression Φ can be further approximated by neglecting the squared terms in the integration variables. From this follows the *Fraunhofer approximation* of the diffraction integral

$$E(x,y;R_0)=\frac{ik}{2\pi}\frac{e^{ikR_0}}{R_0}e^{\frac{ik}{2R_0}(x^2+y^2)}\iint_A E(x',y')e^{\frac{ik}{R_0}(xx'+yy')}dx'dy', \tag{4.73}$$

or using the above mentioned dimensionless coordinates

$$E(\xi,\eta)=i\sqrt{N_{F,x}N_{F,y}}e^{ikR_0}e^{i\pi\left(N_{F,x}\xi^2+N_{F,y}\eta^2\right)}\int_{-1}^{1}\int_{-1}^{1}E(\xi',\eta')e^{2i\pi\left(N_{F,x}\xi\xi'+N_{F,y}\eta\eta'\right)}d\xi'd\eta'. \tag{4.74}$$

Except for the prefactors, Eq. 4.73 or Eq. 4.74 represents the *Fourier transform* of the field distribution at the aperture plane. The Fraunhofer approximation is valid for the Fresnel numbers $N_{F,\,x}$, $N_{F,\,y}\ll 1$, hence for the distances $z\gg a_{x,y}^2\big/\lambda$ from the aperture. In addition, the condition has to be fulfilled that the wavefront is nearly planar at the aperture plane. This condition is fulfilled especially when the light source is located far from the aperture.

4.8.4 Diffraction at the Slit

A commonly occurring situation is the diffraction of an incident, planar wave at a narrow slit. The assumption of a planar wave of constant amplitude over the entire slit is, as a rule, justified by the small slit dimensions in relation to the beam diameter and to the distance of the source. In order to identify the essential characteristics of the diffraction at the slit, it suffices to observe only one transverse coordinate.

The experimental setup for the diffraction at a slit is represented in Fig. 4.33 schematically. The distance to the screen D is very large in relation to the width of the slit a. In this way, the condition $N_F \ll 1$ is fulfilled for the Fraunhofer approximation, and the intensity distribution of the diffraction pattern on the screen can be calculated according to Eq. 4.73. The slit is illuminated from the left using a planar wave of constant amplitude. Furthermore, it is assumed that D is also much larger than the transverse expansion of the diffraction pattern such that the approximations

$$R_0 = \sqrt{D^2 + x^2} \approx D, \sin\theta = \frac{x}{D} \approx \theta \tag{4.75}$$

are always valid. For this reason, in a one-dimensional case Eq. 4.74 becomes

$$E(x;D) = \frac{ik}{2\pi}\frac{e^{ikD}}{D} e^{\frac{ik}{2D}x^2} \int_{-a/2}^{a/2} E_0 e^{\frac{ik}{D}xx'} \mathrm{d}x'. \tag{4.76}$$

E_0 Amplitude of the incident wave
a Width of the slit
D Distance to the screen

The phase of the incident, planar wave is irrelevant for the diffraction and, therefore, omitted. The integration can simply be carried out so that for the field intensity on the screen, it follows:

$$\begin{aligned} E(x;D) &= \frac{ik}{2\pi}\frac{e^{ikD}}{D} e^{\frac{ik}{2D}x^2} E_0 \cdot \frac{D}{ikx} \cdot \left[e^{\frac{ika}{2D}x} - e^{-\frac{ika}{2D}x} \right] \\ &= \frac{ik}{2\pi}\frac{e^{ikD}}{D} e^{\frac{ik}{2D}x^2} E_0 a \cdot \frac{\sin\left(\frac{ka}{2}\theta\right)}{\left(\frac{ka}{2}\theta\right)}. \end{aligned} \tag{4.77}$$

$\theta = x/D$ Angle to optical axis

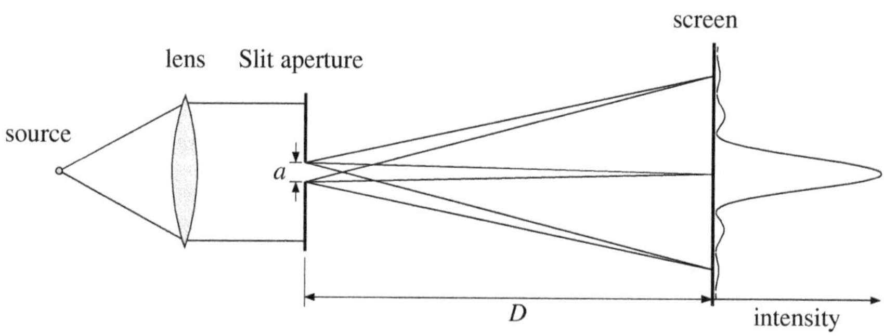

Fig. 4.33 Diffraction at a slit

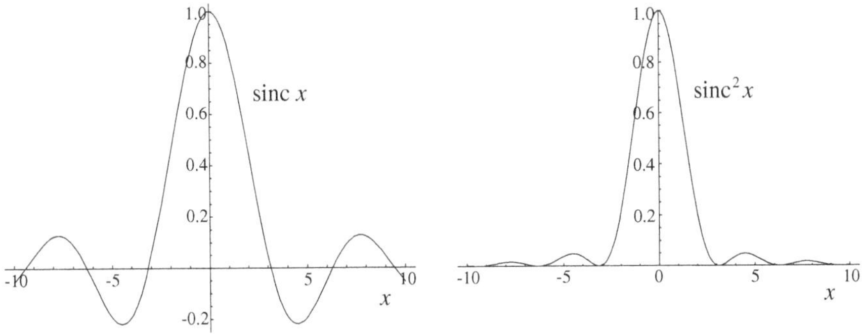

Fig. 4.34 The sinc-function (*left*) and the square of the sinc-function (*right*). The diffraction pattern of the individual slit corresponds to the square of the sinc-function

Except for several phase terms and constant prefactors, the field intensity corresponds to the so-called *sinc-function*:

$$\mathrm{sinc}x = \frac{\sin x}{x}.$$

The intensity is proportional to the square of the absolute value of the field intensity; the diffraction pattern of the individual slit is, therefore, described by (Fig. 4.34)

$$I(\theta) \sim \frac{|E_0|^2 a^2}{\lambda^2 D^2} \cdot \sin c^2\left(\pi \frac{a}{\lambda} \theta\right) \tag{4.78}$$

Diffraction minima are present where the argument of the sinc-function takes on a multiple of π

$$I(\theta) = 0 : \ \pi \frac{a}{\lambda} \theta = n\pi \Leftrightarrow \theta = \frac{n\lambda}{a}, \quad n = \pm 1, \pm 2, \ldots \tag{4.79}$$

This means the narrower the slit is, the further apart the diffraction minima lie, and the wider the diffraction pattern becomes.

4.9 Nonlinear Optics

4.9.1 Maxwell's and Material Equations

Neglecting quantum mechanical effects, emission, propagation, and absorption of electromagnetic radiation can be described by Maxwell's equations. In the field of optics, electrical currents and space charges normally do not occur. In this case Maxwell's equations read:

$$\vec{\nabla} \times \vec{E} = -\frac{\partial \vec{B}}{\partial t} \tag{4.80}$$

$$\vec{\nabla} \times \vec{H} = -\frac{\partial \vec{D}}{\partial t} \tag{4.81}$$

$$\vec{\nabla} \cdot \vec{D} = 0 \tag{4.82}$$

$$\vec{\nabla} \cdot \vec{H} = 0 \tag{4.83}$$

$\vec{E}$ electric field
$\vec{D}$ electric displacement
$\vec{B}$ magnetic field
$\vec{H}$ magnetic induction

The fields $\vec{E}$ and $\vec{D}$ and $\vec{H}$ and $\vec{B}$, respectively, are connected by the relations

$$\vec{D} = \varepsilon_0 \vec{E} + \vec{P} \tag{4.84}$$

$$\overleftarrow{B} = \mu_0 \vec{H} + \vec{M} \tag{4.85}$$

$\vec{P}$ is the macroscopic electric polarization and $\vec{M}$ the macroscopic magnetization. In vacuum $\vec{P}$ and $\vec{M}$ vanish. Except in case of special materials the magnetization can be neglected for optical frequencies. The electric polarization on the other hand is crucial for many optical effects. The polarization in general depends on the material at hand, the light frequency and also on the electric field. The latter is due to the fact that the potential of the oscillating dipoles in the material is not exactly harmonic. Assuming a harmonic oscillator potential is a good approximation for small electric fields.

4.9.1.1 First Order Polarization

In the dipole or local field approximation, i.e., $\vec{P}(\vec{r})$ only depends on the local field $\vec{E}(\vec{r})$ at position $\vec{r}$, the polarization is given by Boyd [1]:

$$\vec{P}^{(1)}(\vec{r},t) = \varepsilon_0 \int_0^\infty \chi^{(1)}(\tau)\vec{E}(\vec{r},t-\tau)\mathrm{d}\tau \tag{4.86}$$

The susceptibility $\chi^{(1)}(\tau)$ is a second-order-tensor. In the general case when $\chi^{(1)}$ has non-vanishing off-diagonal elements, $\vec{P}$ is not parallel to $\vec{E}$ which implies that $\vec{D}$ and $\vec{E}$ are not parallel to each other. This occurs in birefringent crystals. The lower integration limit in Eq. 4.86 is due to causality. The limit could also be extended to $-\infty$ with the susceptibility $\chi^{(1)}(\tau) = 0$ for $\tau < 0$. The Fourier decomposition of the field is given by

$$\vec{E}(\vec{r},t) = \tilde{\vec{E}}(\vec{r},\omega)\exp(-i\omega t)\frac{\mathrm{d}\omega}{2\pi} \tag{4.87}$$

The electric field is a real quantity which implies that $\tilde{\vec{E}}(\vec{r},-\omega) = \tilde{\vec{E}}^{*}(\vec{r},\omega)$. Inserting in Eq. 4.86 yields

$$\vec{P}^{(1)}(\vec{r},t) = \varepsilon_0 \int_0^{\infty} \int_{-\infty}^{\infty} \chi^{(1)}(\tau)\exp(i\omega\tau)d\tau\, \tilde{\vec{E}}(\vec{r},\omega)\exp(-i\omega t)\frac{\mathrm{d}\omega}{2\pi}. \tag{4.88}$$

The integral over τ is the Fourier transform of the susceptibility $\chi^{(1)}$:

$$\vec{P}^{(1)}(\vec{r},t) = \varepsilon_0 \int_{-\infty}^{\infty} \chi^{(1)}(\omega)\tilde{\vec{E}}(\vec{r},\omega)\exp(-i\omega t)\frac{\mathrm{d}\omega}{2\pi}. \tag{4.89}$$

In case of a monochromatic wave this becomes:

$$\vec{P}^{(1)}(\vec{r},t) = \varepsilon_0 \chi^{(1)}(\omega)\tilde{\vec{E}}(\vec{r},\omega) + c.c. \tag{4.90}$$

4.9.1.2 Second-Order Polarization

With increasing electric field strength, the impact of the anharmonicity increases. To account for this nonlinear behavior, the electric displacement is expanded in powers of the electric field. The second order term is given by Shen [2]

$$\vec{P}^{(2)}(\vec{r},t) = \varepsilon_0 \int_0^{\infty}\int_0^{\infty} \chi^{(2)}(\tau_1,\tau_2) : \vec{E}(\vec{r},t-\tau_1)E(\vec{r},t-\tau_2)\,\mathrm{d}\tau_1\mathrm{d}\tau_2\,. \tag{4.91}$$

The colon in the above equation designates a tensor product. Using components this reads:

$$P_i = \sum_j \sum_k \chi_{ijk} E_j E_k. \tag{4.92}$$

Inserting the Fourier transforms of the fields and rearranging yields

$$\vec{P}^{(2)}(\vec{r},t) = \varepsilon_0 \int_0^{\infty}\int_0^{\infty} \chi^{(2)}(\tau_1,\tau_2) : \int_{-\infty}^{\infty} \tilde{\vec{E}}(\vec{r},\omega_1)\exp(-i\omega_1(t-\tau_1))\frac{\mathrm{d}\omega_1}{2\pi}$$
$$\int_{-\infty}^{\infty} \tilde{\vec{E}}(\vec{r},\omega_2)\exp(-i\omega_2(t-\tau_2))\frac{\mathrm{d}\omega_2}{2\pi}\mathrm{d}\tau_1\mathrm{d}\tau_2. \quad (4.93)$$

The integrals over τ_1 and τ_2 are the Fourier transforms of the nonlinear susceptibility

$$\chi^{(2)}(\omega_1+\omega_2;\omega_1,\omega_2) = \int\int \chi^{(2)}(\tau_1,\tau_2)\exp(i(\omega_1\tau_1+\omega_2\tau_2))\mathrm{d}\tau_1\mathrm{d}\tau_2 \quad (4.94)$$

$\chi^{(2)}(\omega_1+\omega_2;\omega_1,\omega_2)$ is in general a complex quantity. Equation 4.93 now becomes

$$\vec{P}^{(2)}(\vec{r},t) = \varepsilon_0 \int_0^{\infty}\int_0^{\infty} \chi^{(2)}(\omega_1+\omega_2;\omega_1,\omega_2) : \tilde{\vec{E}}(\vec{r},\omega_1)\tilde{\vec{E}}(\vec{r},\omega_2)$$
$$\exp(-i(\omega_1+\omega_2)t)\frac{\mathrm{d}\omega_1}{2\pi}\frac{\mathrm{d}\omega_2}{2\pi}. \quad (4.95)$$

In components the second order nonlinear polarization reads

$$P_i^{(2)}(\vec{r},t) = \varepsilon_0 \sum_j\sum_k \int_0^{\infty}\int_0^{\infty} \chi^{(2)}(\omega_1+\omega_2;\omega_1,\omega_2)_{ijk} : \tilde{E}_j(\vec{r},\omega_1)\tilde{E}_k(\vec{r},\omega_1)$$
$$\exp(-i(\omega_1+\omega_2)t)\frac{\mathrm{d}\omega_1}{2\pi}\frac{\mathrm{d}\omega_2}{2\pi}. \quad (4.96)$$

The electric displacement can now be written as

$$\vec{D}^{(2)} = \varepsilon_0\vec{E} + \vec{P}^{(1)} + \vec{P}^{(2)} = \vec{D}^{(1)} + \vec{P}^{(2)}. \quad (4.97)$$

In case of a monochromatic wave, the linear term is given by:

$$\vec{D}^{(1)} = \varepsilon_0(1+\tilde{\chi}(\omega))\tilde{\vec{E}}\exp(-i\omega t) = \varepsilon_0\tilde{\varepsilon}(\omega)\tilde{\vec{E}}\exp(-i\omega t). \quad (4.98)$$

Assuming the two fields in Eq. 4.99 to be monochromatic the nonlinear polarization becomes:

$$P_i^{(2)}(\vec{r},t) = \varepsilon_0 \sum_j\sum_k \chi^{(2)}(\omega_1+\omega_2;\omega_1,\omega_2)_{ijk} : \tilde{E}_j(\vec{r},\omega_1)\tilde{E}_k(\vec{r},\omega_1). \quad (4.99)$$

4.9.2 Wave Equation

From Maxwell's equations the wave equation for the electric field can be deduced

$$\Delta\vec{E}(\vec{r},t) - \nabla(\nabla\cdot\vec{E}(\vec{r},t)) - \mu_0\frac{\partial^2\vec{D}^{(1)}(\vec{r},t)}{\partial t^2} = \mu_0\frac{\partial^2\vec{P}^{(2)}(\vec{r},t)}{\partial t^2} \tag{4.100}$$

4.9.2.1 Separating the Fast Oscillating Factors

The field is now expressed as the product of a slowly varying envelope and a fast oscillating phase

$$\vec{E}(\vec{r},t) = \vec{A}(\vec{r},t)\exp(i(k_0z-\omega t)) + c.c. \tag{4.101}$$

Inserting into the wave equation Eq. 4.100 yields

$$\begin{aligned}\Delta_t\vec{A}(\vec{r},t) + \frac{\partial^2\vec{A}(\vec{r},t)}{\partial z^2} + 2ik_0\frac{\partial\vec{A}(\vec{r},t)}{\partial z}\\ - k_0^2\vec{A}(\vec{r},t) + \mu_0\varepsilon_0\omega_0^2\hat{\varepsilon}\vec{A}(\vec{r},t) = \mu_0\frac{\partial^2\vec{P}(\vec{r},t)}{\partial t^2}\exp(-i(k_0z-\omega_0t))\end{aligned} \tag{4.102}$$

Using Eq. 4.95 the right hand side of Eq. 4.102 becomes

$$\begin{aligned}&\frac{1}{2ik_0}\mu_0\frac{\partial^2P_i^{(2)}}{\partial t^2}(\vec{r},t)\exp(-i(k_0z-\omega_0t))\\ &= \sum_j\sum_k\int_0^\infty\int_0^\infty\frac{i(\omega_1+\omega_2)^2}{2k_0c^2}\chi^{(2)}(\omega_1+\omega_2;\omega_1,\omega_2)_{ijk}\\ &\tilde{E}_j(\vec{r},\omega_1)\tilde{E}_k(\vec{r},\omega_2)\exp(-i(\omega_1+\omega_2-\omega_0)t)\frac{\mathrm{d}\omega_1}{2\pi}\frac{\mathrm{d}\omega_2}{2\pi}\exp(-ik_0z)\end{aligned} \tag{4.103}$$

The Fourier transform of $\tilde{\vec{A}}$ and $\tilde{\vec{E}}$ are related by

$$\tilde{\vec{E}}(\vec{r},\omega) = \tilde{\vec{A}}(\vec{r},\omega-\omega_0)\exp(ik_0z) \tag{4.104}$$

4.9.3 Three Wave Mixing

In the following we assume that the field consists of three distinct parts that oscillate around ω_p, ω_s and ω_i, respectively, with $\omega_s+\omega_i=\omega_p+\Delta\omega$. s is called the signal wave, i the idler and p the pump wave. For the frequency deviation $\Delta\omega$, the condition $\Delta\omega\tau \leq \pi$ has to be met where τ is the interaction time. For further explanation see below.

4.9.3.1 Polarization of the Pump Wave

The polarization of the pump wave depends on the signal and idler waves only. Taking $k_0 = k_p$ in Eq. 4.102 and expressing the fields by their envelopes we get

$$\begin{aligned}
&\tilde{E}_j(\vec{r},\omega_1)\tilde{E}_k(\vec{r},\omega_2)\exp(-ik_pz) = \exp(-ik_pz)\\
&\left[\tilde{A}_j(s;\vec{r},\omega_1-\omega_s)\exp(ik_sz)+\tilde{A}_j(s;\vec{r},\omega_1-\omega_i)\exp(ik_iz)\right]\\
&\left[\tilde{A}_k(s;\vec{r},\omega_2-\omega_s)\exp(ik_sz)+\tilde{A}_k(i;\vec{r},\omega_1-\omega_i)\exp(ik_iz)\right]\\
&=\\
&\tilde{A}_j(s;\vec{r},\omega_1-\omega_s)\tilde{A}_k(s;\vec{r},\omega_2-\omega_s)\exp(i(k_s+k_s-k_p)z)+\\
&\tilde{A}_j(s;\vec{r},\omega_1-\omega_s)\tilde{A}_k(i;\vec{r},\omega_2-\omega_i)\exp(i(k_s+k_i-k_p)z)+\\
&\tilde{A}_j(i;\vec{r},\omega_1-\omega_i)\tilde{A}_k(s;\vec{r},\omega_2-\omega_s)\exp(i(k_i+k_s-k_p)z)+\\
&\tilde{A}_j(i;\vec{r},\omega_1-\omega_i)\tilde{A}_k(i;\vec{r},\omega_2-\omega_i)\exp(i(k_i+k_i-k_p)z)+\\
&\tilde{A}_j(p;\vec{r},\omega_1-\omega_p)\tilde{A}_k(s;\vec{r},\omega_2-\omega_s)\exp(i(k_p+k_s-k_p)z)+\\
&\tilde{A}_j(p;\vec{r},\omega_1-\omega_p)\tilde{A}_k(i;\vec{r},\omega_2-\omega_i)\exp(i(k_p+k_i-k_p)z)
\end{aligned} \tag{4.105}$$

The terms 5 and 6 are fast oscillating so that they do not contribute to the conversion. If $k_s \neq k_i$ also term 1 and 4 are fast oscillating. Thus two terms remain

$$\begin{aligned}
&\tilde{E}_j(\vec{r},\omega_1)\tilde{E}_k(\vec{r},\omega_2)\exp(-ik_pz) = \exp(i(k_s+k_i-k_p)z)\\
&\left[\tilde{A}_j(s;\vec{r},\omega_1-\omega_s)\tilde{A}_k(i;\vec{r},\omega_2-\omega_i)+\tilde{A}_j(i;\vec{r},\omega_1-\omega_i)\tilde{A}_k(s;\vec{r},\omega_2-\omega_s)\right]
\end{aligned} \tag{4.106}$$

Inserting Eq. 4.106 into Eq. 4.103 yields

$$\begin{aligned}
&\frac{1}{2ik_p}\mu_0\frac{\partial^2 P_i^{(2)}}{\partial t^2}(\vec{r},t)\exp(-i(k_pz-\omega_pt)) = \exp(i(k_s+k_i-k_p)z)\\
&\sum_j\sum_k\int_0^\infty\int_0^\infty\frac{i(\omega_1+\omega_2)^2}{2k_pc^2}\chi^{(2)}(\omega_1+\omega_2;\omega_1,\omega_2)_{ijk}\\
&\left[\tilde{A}_j(s;\vec{r},\omega_1-\omega_s)\tilde{A}_k(i;\vec{r},\omega_2-\omega_i)+\tilde{A}_j(i;\vec{r},\omega_1-\omega_i)\tilde{A}_k(s;\vec{r},\omega_2-\omega_s)\right]\\
&\exp(-i(\omega_1+\omega_2-\omega_p)t)\frac{d\omega_1}{2\pi}\frac{d\omega_2}{2\pi}
\end{aligned} \tag{4.107}$$

4.9.3.2 Phase Matching

In order to get high conversion efficiency between the three interacting waves the phase of the factor $\exp(-i(\omega_1+\omega_2-\omega_p)t)$ in the integral must not exceed π during the interaction time because otherwise the conversion direction would change sign. This implies that the field amplitudes $\tilde{A}_j$ must be sharply centered around ω_s and ω_i respectively (see above).

Besides this matching of frequencies, the phase $(k_s+k_i-k_p)z$ has to be small within the interaction length z. This condition is called phase matching condition. In vacuum the phase matching condition would be fulfilled if the frequency matching condition is met. In material this no longer holds and special arrangements have to be used. The most widely used mechanism is index matching. This means that the refractive indices of the three interacting waves are chosen in such a way that the phase matching condition is fulfilled. This can be achieved by exploiting dispersion and birefringence. If for example the s and i waves are polarized along a crystal direction with large refractive index the refractive index of the p wave can be fitted as required by choosing a polarization direction with lower refractive index that acquires the necessary refractive index value at the larger p wave frequency due to normal dispersion.

Another mechanism to meet the phase matching condition is to use so-called quasi phase matching. In this case the nonlinear crystal is periodically poled in such a way that every coherence length:

$$z_{\mathrm{coh}} = \frac{\pi}{k_s+k_i-k_p} \tag{4.108}$$

the nonlinear susceptibility $\chi^{(2)}_{ijk}$ changes its sign. This changes the phase by π so that the dephasing due to phase mismatch is compensated for. The main advantage of quasi phase matching is that all three waves can have equal polarization for which in the general case the coupling coefficient $\chi^{(2)}_{iii}$ is largest.

4.9.3.3 Signal and Idler Waves

Equation 4.107 is the polarization envelope which drives the p-wave. To find the polarization that drives the s-wave and assuming that $\tilde{\vec{E}}(\vec{r},\omega_1)$ is centered around ω_p the field $\tilde{\vec{E}}(\vec{r},\omega_2)$ must be centered around $(-\omega_i)$ in order that the term oscillating in time in the integral does not change considerably within the period of time of interest. With

$$\tilde{\vec{E}}(\vec{r},-\omega_2) = \tilde{\vec{E}}^*(\vec{r},\omega_2) \tag{4.109}$$

the polarization is

$$\frac{1}{2ik_s}\mu_0\frac{\partial^2 P_i^{(2)}}{\partial t^2}(\vec{r},t)\exp(-i(k_s z-\omega_s t))=\exp(i(k_p-k_i-k_s)z)$$
$$\sum_j\sum_k\int_0^\infty\int_0^\infty\frac{i(\omega_1+\omega_2)^2}{2k_pc^2}\chi^{(2)}(\omega_1+\omega_2;\omega_1,\omega_2)_{ijk}$$
$$\left[\tilde{A}_j(s;\vec{r},\omega_1-\omega_p)\tilde{A}_k^*(i;\vec{r},\omega_2-\omega_i)+\tilde{A}_k(p;\vec{r},\omega_2-\omega_p)\tilde{A}_k^*(i;\vec{r},\omega_2-\omega_i)\right]$$
$$\exp(-i(\omega_1+\omega_2-\omega_s)t)\frac{\mathrm{d}\omega_1}{2\pi}\frac{\mathrm{d}\omega_2}{2\pi} \tag{4.110}$$

An equivalent equation holds for the polarization of the idler-wave

$$\frac{1}{2ik_i}\mu_0\frac{\partial^2 P_i^{(2)}}{\partial t^2}(\vec{r},t)\exp(-i(k_i z-\omega_i t))=\exp(i(k_p-k_i-k_s)z)$$
$$\sum_j\sum_k\int_0^\infty\int_0^\infty\frac{i(\omega_1-\omega_2)^2}{2k_pc^2}\chi^{(2)}(\omega_1-\omega_2;\omega_1,-\omega_2)_{ijk}$$
$$\left[\tilde{A}_j(p;\vec{r},\omega_1-\omega_p)\tilde{A}_k^*(s;\vec{r},\omega_2-\omega_s)+\tilde{A}_k(p;\vec{r},\omega_1-\omega_p)\tilde{A}_j^*(s;\vec{r},\omega_2-\omega_s)\right]$$
$$\exp(-i(\omega_1-\omega_2-\omega_i)t)\frac{\mathrm{d}\omega_1}{2\pi}\frac{\mathrm{d}\omega_2}{2\pi} \tag{4.111}$$

4.9.3.4 Walk-off

As already mentioned, $\chi^{(1)}$ in general is a tensor and so $\vec{E}$ is in general not parallel to $\vec{D}$. $\vec{D}$ is perpendicular to the direction of the wave vector $\vec{k}$ and the same holds for $\vec{H}$ so that $\vec{D}\times\vec{H}$ is parallel to $\vec{k}$ but the Poynting vector $\vec{E}\times\vec{H}$ in general is not. This means that the direction of propagation of the phase front and the direction of power flow are in general not parallel. Depending on the propagation direction and the polarization this can cause a lateral off-set between the optical axis and the beam centroid along the propagation direction. This phenomenon is called walk-off. In case of index matching this can cause beams that have different polarization directions to diverge along the interaction length which reduces the conversion efficiency.

References

1. Boyd, Robert W.: Nonlinear Optics. Academic Press, 2003
2. Shen, Yuen-Ron: The Principles of Nonlinear Optics. John Wiley, 1984

Chapter 5
Laser Beams

Typically, the apertures of optical systems are large compared to their wavelength. In most cases, therefore, the ray optics concept represents a sufficiently accurate approximation to describe these optical systems. For small apertures, however, this concept is no longer valid when the wave nature of light predominates and diffraction effects are not negligible.

Nonetheless, diffraction is an important mechanism behind the generation of stationary wave fields in laser resonators. Even though the diffraction effects per round trip in the resonator may be small, the accumulated effect after a large number of round trips is significant. Additionally, a power loss results from diffraction. Indeed, comparably small losses per round trip can already significantly change the beam characteristics of a laser cavity.

Accounting for these diffraction effects requires solving the wave equation without using the ray optics approximation, i.e., taking into account the effects of a finite wavelength. Besides the Fresnel–Kirchhoff diffraction theory, a second model has been established to specifically describe wave fields propagating into a narrow solid angle. This model is called the *Slowly Varying Envelope Approximation*, or *SVE approximation*, and its expression is applicable to wave fields whose envelope only slowly changes in time and space.

Since the SVE approximation is specifically tailored to describe a directed propagation of wave fields into a narrow solid angle, it can be seen as a model of a light ray. Although this model is suitable for rays of "classical" light as well, the following chapter focuses solely on laser beams. To a certain extent this anticipates the detailed discussion of the working principles of the laser and the fundamental characteristics of laser beams. In this chapter, the laser still has to be seen as a "black box" emitting a nearly ideal light beam. This approach is advisable because the beam propagation in SVE approximation is also applied to the description of the wave fields inside of the laser cavity.

R. Poprawe et al., *Tailored Light 1*, RWTHedition,
DOI 10.1007/978-3-642-01234-1_5

5.1 The SVE Approximation

In general, the scalar wave equation

$$\Delta E + k^2 E = 0 \tag{5.1}$$

has an infinite number of solutions. We are looking for solutions describing waves propagating within a narrow solid angle into a specific direction, e.g., along the positive z-axis. The simplest solution fulfilling this condition is a plane wave

$$E = E_0 e^{ikz}. \tag{5.2}$$

However, a plane wave is not suitable to describe a real wave field since it is of infinite, lateral extension. To account for a lateral limitation of the wave field, the following approach is made

$$E(x, y, z) = E_0(x, y, z)e^{ikz}. \tag{5.3}$$

Thus, the amplitude, which is constant for a plane wave, is now a function of the position in space.

Applying this approach to the scalar wave equation yields

$$\frac{\partial^2 E_0}{\partial x^2} + \frac{\partial^2 E_0}{\partial y^2} + \frac{\partial}{\partial z}\left(\frac{\partial E_0}{\partial z} + 2ikE_0\right) = 0. \tag{5.4}$$

This expression is still exact. Now, it is assumed that the amplitude $E_0(x,y,z)$ only changes negligibly within one wavelength

$$\frac{\partial E_0}{\partial z} \ll kE_0 = 2\pi \frac{E_0}{\lambda}. \tag{5.5}$$

This is the so-called *Slowly Varying Envelope* or *SVE approximation.* This assumption is equivalent to the assumption that the wave field as a whole propagates along the z-axis, which has been determined as the optical axis in this case. For this reason, the SVE approximation is also called the *paraxial approximation.* Applying this approximation yields the *wave equation in SVE approximation* or *paraxial wave equation* in Cartesian coordinates

$$\frac{\partial^2 E_0}{\partial x^2} + \frac{\partial^2 E_0}{\partial y^2} + 2ik\frac{\partial E_0}{\partial z} = 0. \tag{5.6}$$

In laser resonators, frequently a rotational symmetric intensity distribution is observed. In cylindrical coordinates,[1] Eq. 5.6 transforms to

$$\frac{\partial^2 E_0}{\partial r^2}+\frac{1}{r}\frac{\partial E_0}{\partial r}+\frac{1}{r^2}\frac{\partial^2 E_0}{\partial \phi^2}+2ik\frac{\partial E_0}{\partial z}=0. \tag{5.7}$$

For rotationally symmetric fields, E_0 does not depend on φ, and Eq. 5.7 simplifies to the paraxial wave equation for rotationally symmetric fields

$$\frac{\partial^2 E_0}{\partial r^2}+\frac{1}{r}\frac{\partial E_0}{\partial r}+2ik\frac{\partial E_0}{\partial z}=0. \tag{5.8}$$

Now, solutions have to be found for Eqs. 5.6 and 5.8. A set of *particular solutions* can be derived analogously to the wave equation in Sect. 2.3. Any linear combination of the particular solutions forms a general solution. In the context of laser beams, the particular solutions are called "*modes of propagation*" or simply "*modes.*" A mode, therefore, is a particular propagation form of a wave field. In general, a real laser beam is a superposition of several modes.

As for the general wave equation, the solutions of the paraxial wave equation in free space are transverse waves, i.e., electrical and magnetic field vectors are perpendicular to the propagation axis. Modes with these characteristics are called transverse electromagnetic modes or *TEM modes*.

5.2 The Gaussian Beam

The lowest-order particular solution of the wave equation in SVE approximation (Eq. 5.8) is the *fundamental mode* or *Gaussian beam*:

$$E(r,z)=E_0\frac{w_0}{w(z)}\mathrm{e}^{-\frac{r^2}{w(z)^2}}\mathrm{e}^{-i\{\omega t-\psi_T(r,z)-\psi_L(z)\}} \tag{5.9}$$

w_0	Beam radius at $z = 0$ (beam waist)
$z_R=\frac{\pi w_0^2}{\lambda}$	Rayleigh length
$w(z)=w_0\sqrt{1+(z/z_R)^2}$	Beam radius in distance z to the beam waist
$R(z)=z\left\{1+(z_R/z)^2\right\}$	Radius of curvature of the phase front
$\psi_T(r,z)=\frac{kr^2}{2R(z)}$	Transverse phase factor
$\psi_L(z)=kz-\arctan\frac{z}{z_R}$	Longitudinal phase factor.

[1] $x = r\cos\varphi$, $y = r\sin\varphi$, $z = z$.

The field distribution of the Gaussian beam comprises an amplitude factor and a phase factor. The amplitude factor determines the lateral intensity distribution of the mode. The phase factor is specifically significant in the context of determining the eigenfrequencies of resonators (see Sect. 5.3.3).

5.2.1 The Amplitude Factor

According to Eq. 5.9, the amplitude factor of the Gaussian beam is

$$A(r,z) = E_0 \frac{w_0}{w(z)} \mathrm{e}^{-\frac{r^2}{w(z)^2}}. \tag{5.10}$$

The amplitude decreases independently of the propagation coordinate z with distance r to the beam axis according to a *Gaussian profile.*[2] The Gaussian beam borrows its name from this characteristic. $w(z)$ is the *beam radius* of the Gaussian beam at the position z, relative to the beam waist at $z = 0$. The beam radius is defined as the distance to the beam axis at which the intensity drops to $1/e$ of its respective value on the axis. The course of $w(z)$ is called the *beam caustic* (Fig 5.1).

The beam radius has its minimum at the *beam waist* at $z = 0$. At $z = z_R$ it yields $w(z_R) = \sqrt{2}w_0$, i.e., the cross section of the beam has doubled. The distance $2z_R$ is, thus, also called the *focal length* of the beam. The focal length is equivalent to the *depth of focus*; it reveals over which distance the beam can be considered to be in focus.

For $z \gg z_R$, the beam radius increases approximately linearly with z. In general, the *far-field divergence angle* of the beam is defined as

$$\Theta = \lim_{z\to\infty} \frac{w(z)}{z}. \tag{5.11}$$

With this definition,

$$\Theta \equiv \Theta_{00} = \frac{w_0}{z_R} = \frac{\lambda}{\pi w_0} \tag{5.12}$$

Θ_{00} Diffraction-limited divergence angle

results for the far-field divergence angle of the Gaussian beam. The Gaussian beam exhibits the minimum divergence angle physically possible for a beam with a given waist radius. Therefore, Θ_{00} is designated as the *diffraction-limited divergence angle* or simply the *diffraction limit.*

---

[2] A Gaussian or normal distribution $f(x)$ is given by $f(x) = 1/\left(\sqrt{2\pi}b\right)\exp\left\{-(x-a)^2/2b^2\right\}$. a and b are the parameters of the distribution. The maximum and symmetry center lie at the position $x = a$. b is the distance of the inflection point from the symmetry center.

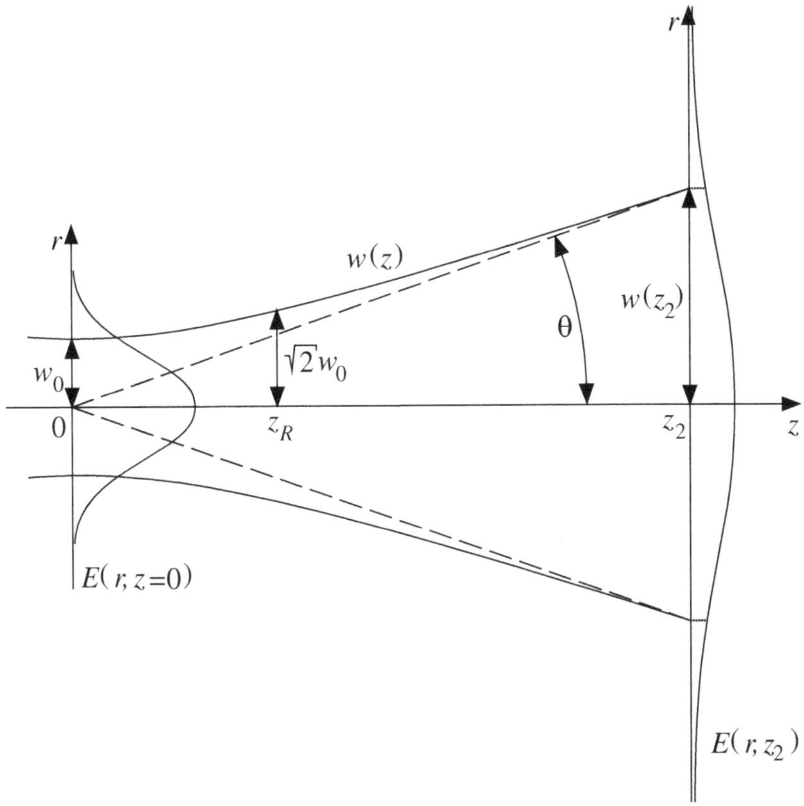

Fig. 5.1 The field amplitude of the Gaussian beam as a function of the propagation coordinate z

5.2.2 *The Phase Factor*

The phase factor (Fig. 5.2)

$$P(r,z) = \mathrm{e}^{-i(\omega t - \psi_T - \psi_L)} \tag{5.13}$$

from Eq. 5.9 contains a transverse contribution ψ_T and a longitudinal contribution ψ_L. The longitudinal term describes the phase oscillation of the wave propagating in $+z$ direction and is, with the exception of a small shift factor,[3] equal to the phase factor of a plane wave. Planes with constant transverse phase, $\psi_T = const.$, have the shape of

[3]The *Guoy Shift* $\arctan(z/z_R)$ of the phase exhibits different values for the higher-order modes. This leads to slightly different eigenfrequencies in spherical resonators (cf. Sect. 6.3.3).

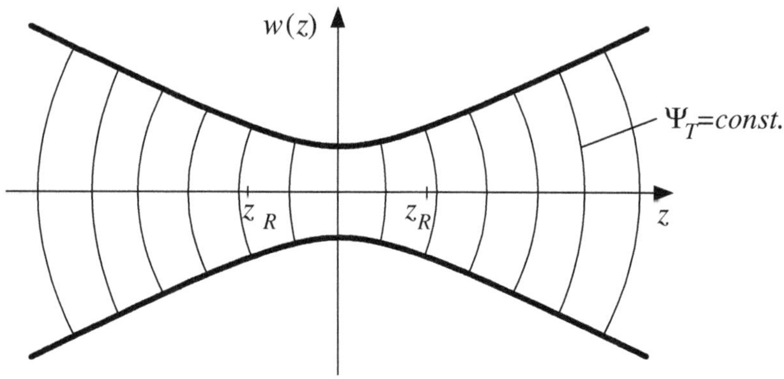

Fig. 5.2 Planes of constant phase of the Gaussian beam

paraboloids of revolution. In proximity to the beam axis they can be approximated by spheres. Their radius of curvature is given by

$$\left(\frac{\partial^2}{\partial r^2}\frac{\psi_T(r,z)}{k}\right)^{-1} = R(z).$$

In both limits

$$z \to \infty : R(z) = \infty \text{ and } z \to 0 : R(z) = \infty \tag{5.14}$$

the radius of curvature approaches infinity, i.e., the planes of constant phase are flat at the beam waist and at an infinite distance to the waist. The radius of curvature exhibits its minimum at $z = z_R$:

$$\frac{\mathrm{d}R}{\mathrm{d}z} = 1 - \frac{z_R^2}{z^2} = 0 \Rightarrow z = \pm z_R, \quad R(z_R) = 2z_R. \tag{5.15}$$

At these positions the phase planes thus have their maximum curvature.

5.2.3 The Intensity Distribution of the Gaussian Beam

The intensity is proportional to the square of the field amplitude (cf. Eq. 3.72), (Fig. 5.3)

$$I = \frac{\varepsilon_0 \varepsilon c}{2} E E^*, \tag{5.16}$$

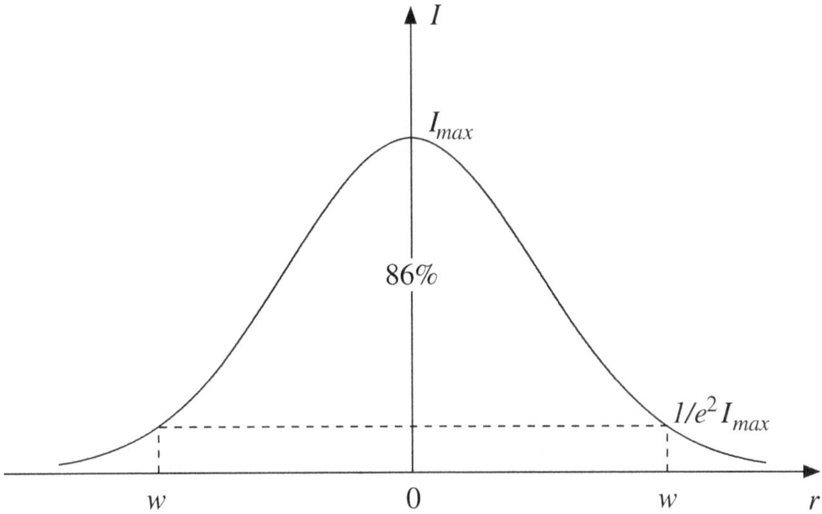

Fig. 5.3 Intensity distribution of the Gaussian beam

resulting in the following intensity distribution for the Gaussian beam:

$$I(r,z) = \frac{\varepsilon_0 \varepsilon c}{2} A(r,z)^2 = \frac{\varepsilon_0 \varepsilon c}{2} E_0^2 \frac{w_0^2}{w(z)^2} \mathrm{e}^{-\frac{2r^2}{w(z)^2}} \equiv I_0(z) \mathrm{e}^{-\frac{2r^2}{w(z)^2}}. \tag{5.17}$$

As the field amplitude, the intensity radially exhibits a Gaussian distribution that broadens while propagating from the beam waist along the z-axis, in accordance with the beam caustic $w(z)$. The beam radius is marked by the intensity's value dropping to $1/e^2$ of its value on the beam axis.

The optical power contained in the beam is calculated by integrating the radial intensity distribution, resulting in a relationship between the peak intensity $I_0(z)$ on the beam axis and the optical power P:

$$P = 2\pi \int_0^\infty I(z,r) r \, \mathrm{d}r = \frac{\pi}{2} w(z)^2 I_0(z) \Leftrightarrow I_0(z) = \frac{2P}{\pi w(z)^2}. \tag{5.18}$$

Since no absorption occurs during free propagation, the power contained in the beam is independent of the propagation coordinate z. From this it follows that the intensity on the beam axis decreases with increasing z and, thus, increasing beam cross section; asymptotic for large z it decreases proportionally to $1/z^2$.

The area enclosed by the beam radius $w(z)$ contains 86 % of the total beam power

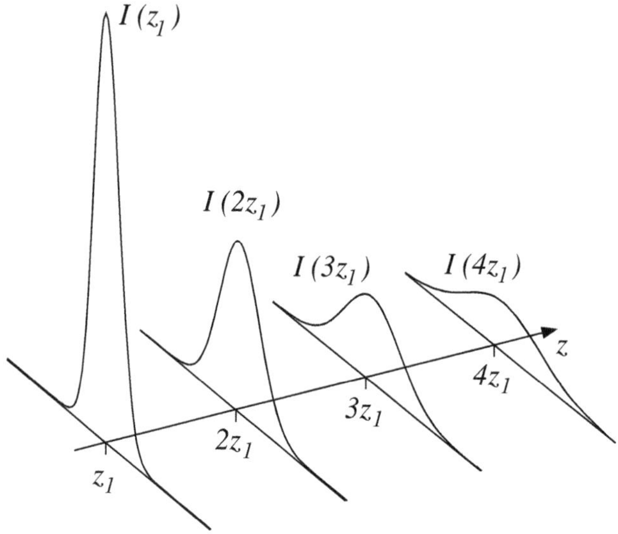

Fig. 5.4 Development of the intensity distribution of the Gaussian beam with increasing propagation coordinate z

$$\int_0^{w(z)} I(r,z) r\,\mathrm{d}r = \left(1 - \frac{1}{e^2}\right) \cdot P \cong 0.86 \cdot P \equiv P_w.$$

Reciprocally, this means still 14 % of the power is transported outside of the beam radius as it is defined here; this is important to consider when designing optical systems or beam apertures (Fig. 5.4).

5.3 Higher-Order Modes

The *higher-order modes*, thus the remaining particular solutions of the wave equation in SVE approximation, are represented by sets of polynomials, depending on the chosen coordinate system. Typically, the choice of the coordinate system reflects the geometry of the beam-defining apertures.

5.3.1 The Hermite-Gaussian Modes

For the paraxial wave equation in Cartesian coordinates, Eq. 5.6, the full set of particular solutions is given by the Hermite-Gaussian modes:

$$E_{mn}(x,y,z) = E_0 H_m\left(\sqrt{2}\frac{x}{w(z)}\right) H_n\left(\sqrt{2}\frac{y}{w(z)}\right)\frac{w_0}{w(z)} \mathrm{e}^{-\frac{x^2+y^2}{w(z)^2}} \mathrm{e}^{i\psi}, \quad m,n = 0,1,2,\ldots$$
$$\psi = kz - (m+n+1)\arctan\frac{z}{z_R} + \frac{z(x^2+y^2)}{z_R w(z)^2}, \tag{5.19}$$

where $H_n(u)$ are the *Hermite polynomials*, and m and n are the order of the mode in x and y directions, respectively. The Hermite polynomials are defined by the differential equation

$$H_n(u) = (-1)^n \mathrm{e}^{u^2} \frac{\mathrm{d}^n}{\mathrm{d}u^n} \mathrm{e}^{-u^2}. \tag{5.20}$$

Thus, the four lowest order Hermite polynomials are

$$\begin{array}{ll} H_0(u) = 1 & H_1(u) = 2u \\ H_2(u) = 4u^2 - 2 & H_3(u) = 8u^3 - 12u. \end{array}$$

From these, e.g., the following Hermite-Gaussian modes can be derived:

$$\begin{array}{ll} E_{00} = E_0 \frac{w_0}{w(z)} \mathrm{e}^{-\frac{x^2+y^2}{w(z)^2}} \mathrm{e}^{i\psi_{00}}, & \psi_{00} = kz - \arctan\frac{z}{z_R} + \frac{z(x^2+y^2)}{z_R w(z)^2} \\ E_{10} = E_0 \frac{w_0}{w(z)} 2\sqrt{2}\frac{x}{w(z)} \mathrm{e}^{-\frac{x^2+y^2}{w(z)^2}} \mathrm{e}^{i\psi_{10}}, & \psi_{10} = kz - 2\arctan\frac{z}{z_R} + \frac{z(x^2+y^2)}{z_R w(z)^2} \\ E_{20} = E_0 \frac{w_0}{w(z)} \left(8\frac{x^2}{w(z)^2} - 2\right) \mathrm{e}^{-\frac{x^2+y^2}{w(z)^2}} \mathrm{e}^{i\psi_{20}}, & \psi_{20} = kz - 3\arctan\frac{z}{z_R} + \frac{z(x^2+y^2)}{z_R w(z)^2}. \end{array} \tag{5.21}$$

E_{00} is the field distribution of the Gaussian beam or fundamental mode; the other field distributions represent higher-order modes. In Fig. 5.5 the field distributions of these modes as well as their construction from the superposition of the Gaussian distribution with the respective Hermite polynomial are represented. As for the Gaussian beam, for all higher-order modes, the fundamental field distribution remains unchanged during propagation, while the width of the distribution increases.

The Hermite-Gaussian modes are commonly designated as *TEM_{nm} modes*. The indices m and n indicate the number of zeros of the respective intensity distribution in x- and y- axis and are called the order of the mode. Intensity distributions of the lowest-order Hermite-Gaussian modes are shown in Fig. 5.6.

The Hermite-Gaussian polynomials form a complete and orthogonal set of functions. This means that any field distribution can be constructed by a superposition of Hermite-Gaussian modes: the general solution of the paraxial wave equation (Eq. 5.6) thus can be written in the form

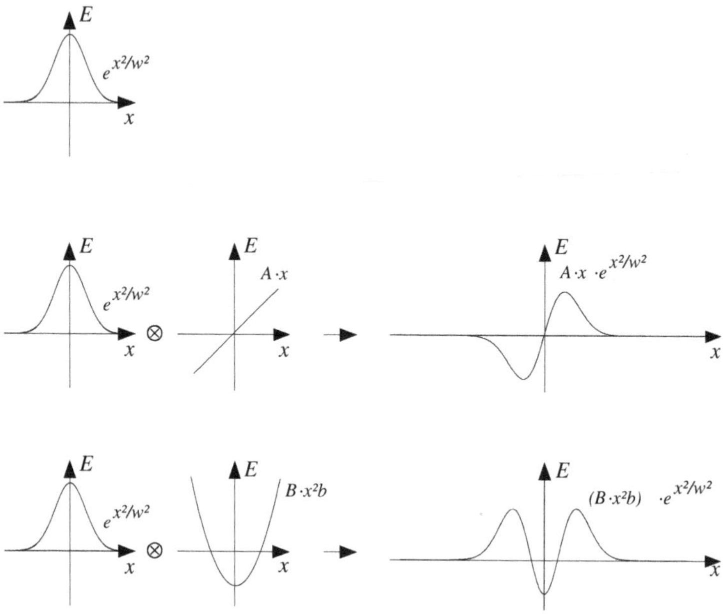

Fig. 5.5 One-dimensional representation of the field amplitudes of the Hermite-Gaussian modes TEM_{00}, TEM_{10}, and TEM_{20}, and their construction by superimposing the individual terms (see. Eq. 5.19)

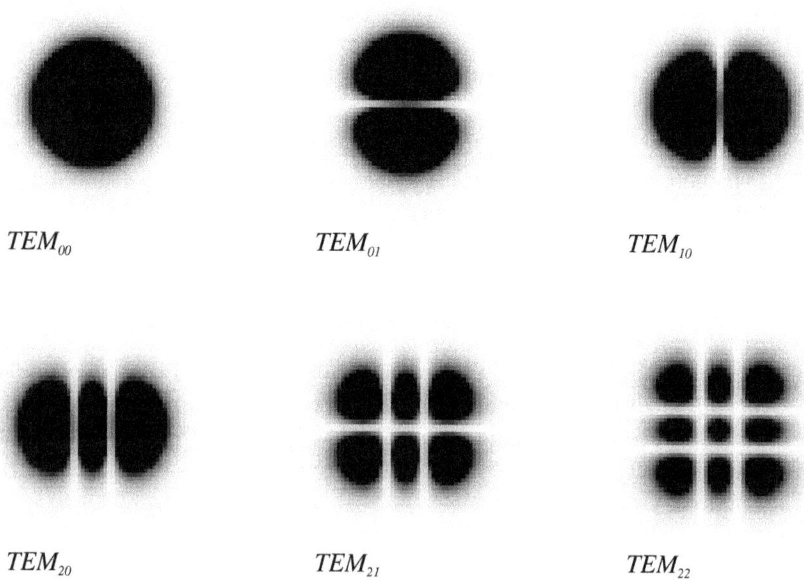

Fig. 5.6 Intensity distributions (cross sections) of several Hermite-Gaussian modes TEM_{mn}

$$E(x,y,z) = \sum_m \sum_n a_{mn} E_{mn}(x,y,z). \tag{5.22}$$

The beam waist radius w_0 represents an additional parameter of the solution. In general, it is defined by the respective geometry, e.g., by the dimensions of beam apertures.

5.3.2 The Laguerre-Gaussian Modes

Similarly, the particular solutions of the paraxial wave equation in cylindrical coordinates (Eq. 5.7) are given by the *Laguerre-Gaussian modes*:

$$\begin{aligned} E_p^l(r,\phi,z) &= E_0 \cdot \left(\frac{\sqrt{2}r}{w(z)}\right)^l L_p^l\left(2\frac{r^2}{w(z)^2}\right) \mathrm{e}^{-\frac{r^2}{w(z)^2}} \mathrm{e}^{i\psi} \begin{cases} \cos(l\phi) \\ \sin(l\phi) \end{cases}, \quad p,l = 0,1,2,\ldots \\ \psi &= kz - (p+l+1)\arctan\left(\frac{z}{z_R}\right) + \frac{zr^2}{z_R w(z)^2}. \end{aligned} \tag{5.23}$$

In this case, two families of solutions exist: the sine and cosine solutions are rotated by 90° to each other. The functions $L_p^l(u)$ are the *Laguerre polynomials*, defined by the differential equation

$$L_p^l(u) = \mathrm{e}^u u^{-l} \frac{\mathrm{d}^p}{\mathrm{d}u^p}\left(\mathrm{e}^{-u} u^{p+l}\right).$$

The first four polynomials are

$$\begin{aligned} L_0^l(u) &= 1 \\ L_1^l(u) &= l+1-u \\ L_2^l(u) &= \frac{1}{2}(l+1)(l+2) - (l+2)u + \frac{1}{2}u^2 \\ L_3^l(u) &= \frac{1}{6}(l+1)(l+2)(l+3) - \frac{1}{2}(l+2)(l+3)u + \frac{1}{2}(l+3)u^2 - \frac{1}{6}u^3. \end{aligned} \tag{5.24}$$

Again E_0^0 is the field distribution of the Gaussian beam; the higher-order modes are obtained for p or l larger than zero. The Laguerre-Gaussian modes are designated by TEM_p^l, with p and l indicating the number of radial and azimuthal zeros, respectively. The intensity distributions of exemplary Laguerre-Gaussian modes are represented in Fig. 5.7.

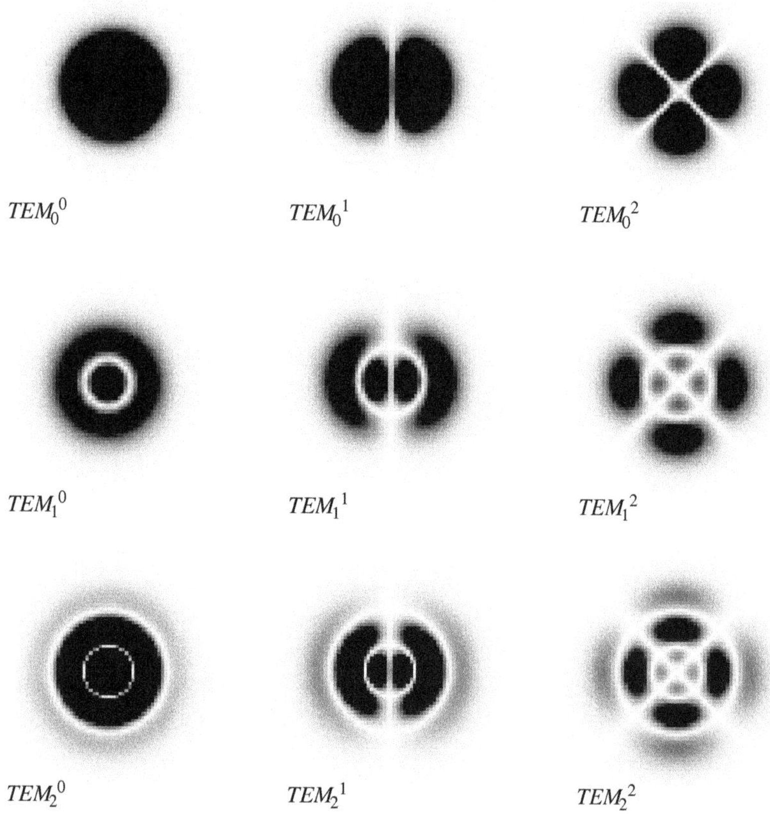

Fig. 5.7 Intensity distributions of Laguerre-Gaussian modes TEM_p^l

Again, the Laguerre polynomials represent a complete and orthogonal set of functions, and the general solution of the paraxial wave equation (Eq. 5.7) is also a linear combination of Laguerre-Gaussian modes.

5.3.3 Doughnut Modes

A special mode type results if the sine and cosine variants of the Laguerre-Gaussian modes are equally present, as can be expected for a fully rotationally symmetric layout. By superposition of the sine and cosine variants with equal amplitude, a set of ring-shaped modes is generated. Consequently, these modes are referred to as *Doughnut modes*:

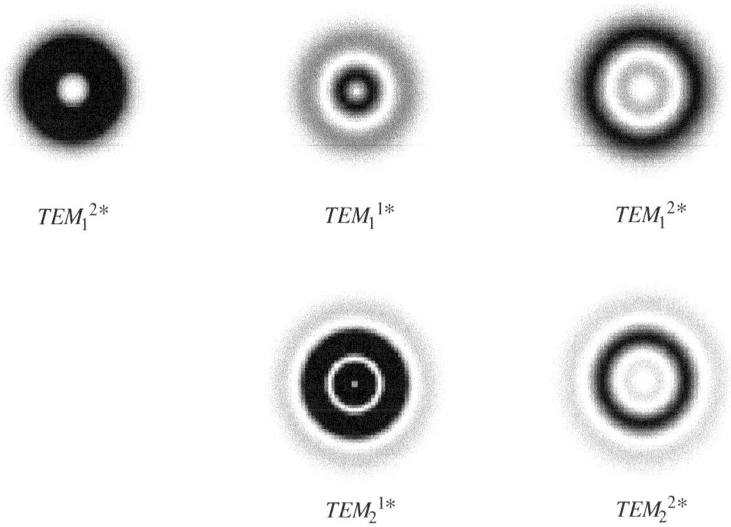

Fig. 5.8 Intensity distributions of some exemplary Doughnut modes TEM_p^l*

$$E_p^{l*}(r,z) = E_0\left(\frac{\sqrt{2}r}{w(z)^2}\right)^l L_p^l\left(2\frac{r^2}{w(z)^2}\right)e^{-\frac{r^2}{w(z)^2}}e^{i\psi}. \tag{5.25}$$

Commonly, for the Doughnut modes the same notation is used as for Laguerre-Gaussian modes, with a star as a differentiator. Figure 5.8 illustrates the intensity distributions of several Doughnut modes.

5.3.4 The Beam Radius of Higher-Order Modes

For the Gaussian beam, the beam waist radius is defined by the decrease of the field amplitude to $1/e$ of its maximum value on the beam axis, and thus, due to the nature of the Gaussian distribution, is given by w_0. For the higher-order modes this correlation is not valid anymore: the multiplication with the Hermite or Laguerre polynomials changes the beam radius, depending on the order of the mode. For the higher-order modes, therefore, w_0 no longer represents the beam waist radius, but is merely a mathematical parameter defining the field distribution.

Furthermore, defining the beam radius based on the $1/e$ drop of the amplitude is not sufficient anymore, because higher-order modes can exhibit more than one radial position at which this criterion is fulfilled. To avoid this ambiguity, the beam radius for the higher order modes is defined as the distance from the beam axis to the outmost inflection point of the field distribution. Since the higher-order

Hermite-Gaussian modes are generally not rotationally symmetric, their beam diameter can be different in the x- and y- directions.[4]

Inflection points of a function are marked by zeros of its second derivative. To simplify the calculation $z = 0$ is chosen. For the Hermite-Gaussian modes, this leads to

$$z = 0 \Rightarrow u = \frac{\sqrt{2}x}{w_0}, \quad E_m \sim H_m(u)\mathrm{e}^{-\frac{u^2}{2}},$$

and the derivatives are

$$E'_m \sim \left\{H'_m(u) - uH_m(u)\right\}\mathrm{e}^{-\frac{u^2}{2}}$$
$$E''_m \sim \left\{\left(u^2 - 1\right)H_m(u) - 2uH'_m(u) + H''_m(u)\right\}\mathrm{e}^{-\frac{u^2}{2}}.$$

With the condition $E''_m = 0$ and the differential equation defining the Hermite polynomials[5]

$$H''_m(u) - 2uH'_m(u) + 2mH_m(u) = 0,$$

the conditional equation for the inflection points u_w follows:

$$H_m(u_w)\mathrm{e}^{-\frac{u_w^2}{2}}\left(u_w^2 - 1 - 2m\right) = 0. \tag{5.26}$$

Obviously, one inflection point is at

$$u_w = \sqrt{2m+1}, \tag{5.27}$$

and further inflection points are given by the zeros of the Hermite polynomial. Necessarily, the zeros are located within the field distribution; thus, with Eq. 5.27 the outmost inflection point is already determined, and the beam radius of the Hermite-Gaussian modes in x- and y- directions is given by

$$w_m = w_0\sqrt{m + \frac{1}{2}} \quad \text{und} \quad w_n = w_0\sqrt{n + \frac{1}{2}}. \tag{5.28}$$

Similarly, the beam radius of the Laguerre-Gaussian modes can be derived to be

$$w_p^l = w_0\sqrt{2p + l + \frac{1}{2}}. \tag{5.29}$$

---

[4]The Gaussian beam can also have different dimensions in both x- and y- directions, when w_0 in the x- and y- directions takes on different values; it is then elliptically distorted.

[5]The solution structure given in Eq. 5.20 follows from this differential equation.

With the general expression

$$w = w_0 b \tag{5.30}$$

W Beam radius of higher-order mode

for the beam radius of the higher-order modes, the far-field divergence angle becomes

$$\Theta = \lim_{z\to\infty} \frac{w(z)}{z} = \frac{w}{z_R} = b\frac{\lambda}{\pi w_0} = b\Theta_{00} \tag{5.31}$$

Θ_{00} Far-field divergence angle of the Gaussian beam
Θ Far-field divergence of the higher-order mode.

Since $b > 1$ for all higher-order modes, beam radius and far-field divergence of all higher-order modes, and any other beam constructed by a superposition of these modes, are larger for a given w_0 than for the Gaussian beam.[6] This distinguishes the Gaussian beam from all other intensity distributions.

A disadvantage of this definition of the beam radius is that it does not converge to the beam radius w_0 of the Gaussian beam for $m = n = 0$ or $p = l = 0$, respectively. For this reason, sometimes a modified definition is chosen:

$$\begin{array}{ll} b = \sqrt{m+1} \text{ bzw. } \; b = \sqrt{n+1} & \text{(Hermite-Gaussian modes)} \\ b = \sqrt{2p+l+1} & \text{(Laguerre-Gaussian modes).} \end{array} \tag{5.32}$$

With this modification the beam radius of the Gaussian beam is reproduced consistently; however, now the beam radii for the higher-order modes deviate slightly from the mathematically consistent definition (Fig. 5.9).

A pragmatic definition, and one more relevant for practical use, of the beam radius is based on the measurement of the enclosed power: The beam radius is given by the radius of an aperture allowing 86 % of the total beam power to pass. This definition is very convenient for the experimental characterization of laser beams. However, it is not suitable for a theoretical description or to predict the propagation characteristics of the beam, since it does not establish a relation between the beam radius, on one hand, and the phase distribution or the diffraction characteristics, on the other.

---

[6] Since the definition $z_R = \pi w_0^2/\lambda$ also remains valid, unchanged, for the higher modes, modes with the same Rayleigh lengths are compared here.

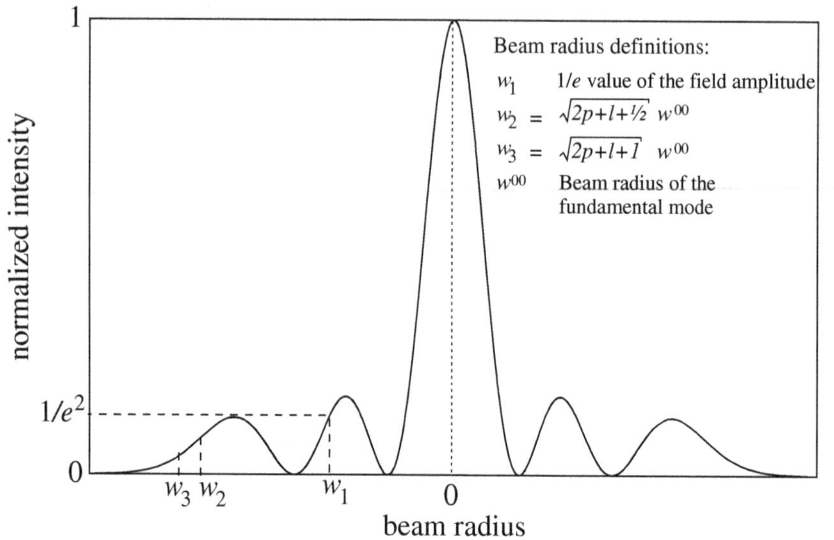

Fig. 5.9 A comparison of the three different definitions of the beam radius of higher order modes

5.4 Real Laser Beams and Beam Quality

Real laser beams only exhibit a Gaussian shape under ideal conditions. Generally, a real laser beam can be described as a superposition of several modes of different orders:

$$E(x,y;z) = \sum_m \sum_n a_{mn} E_{mn}(x,y;z) \text{ or } E(x,y;z) = \sum_p \sum_l a_{pl} E_p^l(x,y;z). \quad (5.33)$$

Since the superposition is linear, the propagation rules for individual modes can be applied to the superposition of modes as well. Again, beam radius and divergence angle represent the main parameters of the beam.

The beam radius and the divergence can be obtained experimentally by measuring the beam caustic. For this measurement the intensity profile of the beam is measured at different locations z along the propagation axis, and the beam radius at these positions is calculated. By fitting the theoretical shape of the caustic $w(z)$ to the experimentally obtained caustic $w_{\text{real}}(z)$, the beam waist radius w_{real}, the Rayleigh length $z_{R,\text{real}}$, and the divergence angle Θ_{real} are derived. These can be related to the parameters of the Gaussian beam, exactly as it was done for the higher order modes in the last section (cf. Eqs. 5.30 and 5.31):

$$w_{\text{real}} = M \cdot w_0, \quad \Theta_{\text{real}} = M \cdot \Theta_{00} \quad \text{with } z_{R,\text{real}} = z_R,$$

where M is now used instead of the coefficient b. Here, the real beam is compared to a Gaussian beam with the same Rayleigh length. The comparison of beams with the same beam waist radius instead leads to

$$
\begin{aligned}
w_{\text{real}} = w_{00} \quad &\Rightarrow \quad w_0 = \tfrac{w_{00}}{M},\, z_{R,\text{real}} = \tfrac{\pi w_0^2}{\lambda} \\
&\Rightarrow \quad \Theta_{real} \equiv \tfrac{w_{\text{real}}}{z_{R,\text{real}}} = w_{00}\left(\tfrac{\pi w_{00}^2}{\lambda M^2}\right)^{-1} = \tfrac{\lambda}{\pi w_{00}} M^2 = \Theta_{00} M^2.
\end{aligned}
$$

For the Gaussian beam $M^2 = 1$, and the divergence angle corresponds to the diffraction limit. For all other beams $M^2 > 1$ holds, and, therefore, the divergence angle of all real beams is larger than the diffraction limit by the factor M^2. The factor M^2 is called *beam propagation factor* and is of great practical importance. This factor can be calculated from the measured beam waist radius w_{real} and the measured divergence angle Θ_{real}:

$$
w_{\text{real}} \Theta_{\text{real}} = \frac{w_{\text{real}}^2}{z_{R,\text{real}}} = \frac{\lambda}{\pi} M^2. \tag{5.34}
$$

The so-called *beam parameter product* $w_{\text{real}}\Theta_{\text{real}}$ is a characteristic constant value for each beam: focusing or defocusing the beam with a lens system always changes beam waist radius and beam divergence reciprocally, so that the beam parameter product remains unchanged.[7]

The beam propagation factor M^2 represents the beam parameter product, normalized to the value of the Gaussian beam. The smaller the beam propagation factor is, the smaller the divergence angle is for a beam with given beam waist radius. The Gaussian beam represents the ideal case with $M^2 = 1$ and the smallest possible divergence angle; in this respect it can, therefore, be considered as the beam with the optimum *beam quality*. Since it is more common for quality indices to increase with increasing quality, in this context the *beam quality index* or *normalized beam quality* is defined as

$$
K = \frac{1}{M^2} = \frac{w_{00}\Theta_{00}}{w_{\text{real}}\Theta_{\text{real}}} = \frac{\lambda/\pi}{w_{\text{real}}\Theta_{\text{real}}}. \tag{5.35}
$$

The beam quality index is one for the Gaussian beam and less than one for all other beams. The smaller the K is, the farther away the beam is from the diffraction limit. A high beam quality means

- a small divergence angle for a given beam waist radius, and
- a small focus radius achieved with a given optical system.

[7]This is valid under the assumption of ideal lenses, transforming the beam without aberrations or optical disturbances. Any disturbance of the beam's phase front leads to an increasing beam parameter product.

Since K (or M^2) is a constant value for any particular beam that is not changed by optical transformations, the beam quality index or the beam propagation factor can not only be considered as a quality measure for the beam itself, but also of the beam source: K (or M^2) is one of the most important characteristic values of a laser beam source. In Fig. 5.10 the achievable beam quality as a function of the output power is shown for different laser beam sources. As a comparison, the normalized beam qualities and the beam propagation factors for some Laguerre-Gaussian modes are summarized in Table 5.1.

5.5 Transformation of Gaussian Beams

An important part of the description of beam propagation in the SVA approximation is the representation of beam transformations by optical elements: in general, the laser beam's characteristics have to be adapted to the specific application by transforming the beam using optical elements such as lenses, prisms, or mirrors. In this section, the Gaussian beam serves as a model representing paraxial beams in general: the general transformation rules are the same for all paraxial beams. For nondiffraction limited beams only the respective scaling of beam radius, divergence, or Rayleigh length with the beam propagation factor has to be considered (see Sects. 5.3.4 and 5.4).

The free propagation of the Gaussian beam is described by the dependence of its characteristics on the propagation coordinate z in Eq. 5.9. The beam widens as a consequence of diffraction, but the shape of its intensity distribution remains unchanged. Only two parameters are required to fully define the Gaussian beam at an arbitrary position, e.g.,

- the Rayleigh length z_R and the distance z to the beam waist position or
- the local beam radius $w(z)$ and the local phase front curvature $R(z)$.

Combining these two parameters in the *complex beam parameter* q leads to a representation of the Gaussian beam that is favorable for the description of its transformation behavior:

$$q = z + iz_R \text{ or } \frac{1}{q} = \frac{1}{R(z)} - \frac{i\lambda}{\pi w(z)^2}. \tag{5.36}$$

With the complex beam parameter the field distribution of the Gaussian beam becomes

$$E(r,z) = E_0 i \frac{z_R}{q} e^{-i\frac{kr^2}{2q}}. \tag{5.37}$$

Since q is based on either one of the above-mentioned sets of two parameters, the parameter q itself is sufficient to completely define the beam.

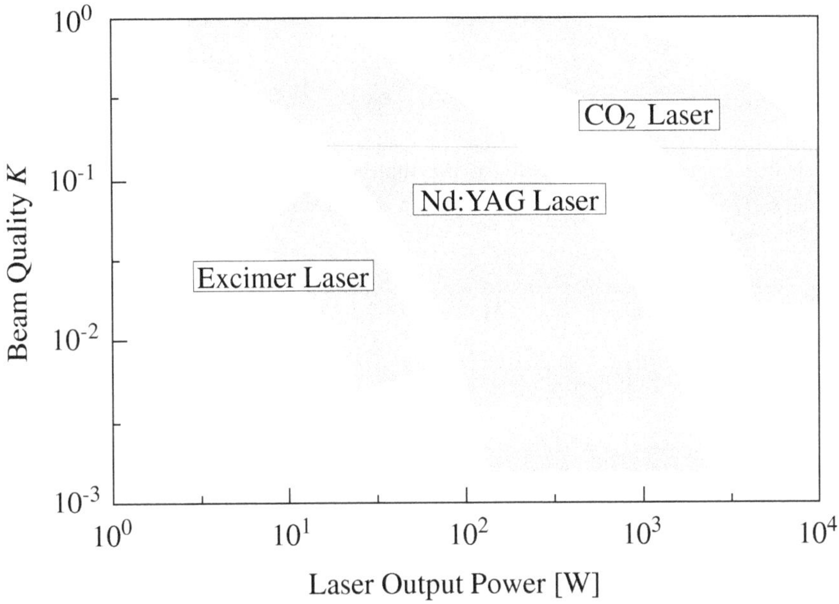

Fig. 5.10 Beam quality of commercially available laser systems (SSL: solid-state laser, HPDL: high-power diode laser). The wavelengths of the respective lasers are: CO_2: 10.6 μm, SSL (Nd: YAG): 1.064 μm, HPDL (GaAlAs): 0.8 μm, Excimer (KrF): 0.248 μm

Table 5.1 The normalized beam quality K and the beam propagation factor M^2 of some Laguerre-Gaussian modes

Mode	K	M^2
TEM_0^0	1	1
TEM_0^1	0.5	2
TEM_0^2	0.33	3
TEM_1^0	0.33	3
TEM_2^0	0.2	5
TEM_1^1	0.25	4
TEM_2^1	0.167	6

5.5.1 The ABCD Law

The most important feature of the complex beam parameter is that it leads to a simple representation of the transformation of Gaussian beams by optical elements: it can be shown that q follows the so-called *ABCD law* for beam transformation:

$$q = \frac{Aq_0 + B}{Cq_0 + D} \tag{5.38}$$

q_0, q Complex beam parameter before and after transformation.

Here, *A*, *B*, *C*, and *D* are elements of the beam transfer matrix **M**,

$$\mathbf{M} = \begin{bmatrix} A & B \\ C & D \end{bmatrix},$$

which has already been defined for the matrix representation of ray optics. For example, the propagation of the beam by a distance *d* is represented by the beam transfer matrix

$$\mathbf{M}_P = \begin{bmatrix} 1 & d \\ 0 & 1 \end{bmatrix}, \tag{5.39}$$

so that from Eqs. 5.38 and 5.36

$$q = q_0 + d,\ q_0 = z_0 + iz_R,\ q = z + iz_R \Rightarrow z = z_0 + d \tag{5.40}$$

follows. In analogy to the matrix representation of ray optics, the transfer matrix for a system consisting of several subsequent optics is obtained by multiplying the matrices of the single optical elements:

$$\mathbf{M} = \mathbf{M}_n \mathbf{M}_{n-1} \cdots \mathbf{M}_2 \mathbf{M}_1. \tag{5.41}$$

The shape of the Gaussian beam remains unchanged under transformations according to the ABCD law. This is a characteristic of all Hermite-Gaussian and Laguerre-Gaussian modes. Therefore, as long as a beam can be modeled as a linear combination of Hermite-Gaussian or Laguerre-Gaussian modes, it transforms according to the ABCD law as well.

5.5.2 Focusing of a Gaussian Beam by a Thin Lens

The beam transfer matrix for a thin lens with focal length *f* is

$$\mathbf{M}_f = \begin{bmatrix} 1 & 0 \\ -\frac{1}{f} & 1 \end{bmatrix}. \tag{5.42}$$

In order to describe the transformation of a Gaussian beam by a thin lens, also the propagation distance from the beam waist to the lens and an additional propagation distance z' to reach a position behind the lens have to be considered:

$$\mathbf{M} = \mathbf{M}_P(z') \mathbf{M}_f \mathbf{M}_P(z_0) \tag{5.43}$$

z_0 Distance between beam waist and lens position
f Focal length of the lens
z' Position behind the lens.

The full transfer matrix thus is

$$\mathbf{M} = \begin{bmatrix} 1 - \frac{z'}{f} & z' + z_0 - \frac{z' z_0}{f} \\ -\frac{1}{f} & 1 - \frac{z_0}{f} \end{bmatrix}, \tag{5.44}$$

and

$$q_f = \frac{A q_0 + B}{C q_0 + D} = \frac{i z_R \left(1 - \frac{z'}{f}\right) + \left(z' + z_0 - \frac{z' z_0}{f}\right)}{-\frac{i z_R}{f} + \left(1 - \frac{z_0}{f}\right)}. \tag{5.45}$$

results for the complex beam parameter of the focused beam q_f. Here, for the beam waist of the unfocused beam, the position $z = 0$ is chosen; thus, the beam parameter of the unfocused beam simplifies to $q_0 = iz_R$. The beam waist position of the focused beam at $z = z_{0f}$ is marked by a minimum beam radius $w_f(z_{0f})$ and a plain phase front, $R_f(z_{0f}) = \infty$.

With Eq. 5.36 it, therefore, follows that

$$\frac{1}{R_f(z_{0f})} = \Re\left(\frac{1}{q_f(z_{0f})}\right) = 0. \tag{5.46}$$

Multiplying the reciprocal of q_f with the conjugate of the denominator to separate the real and imaginary parts results in (Fig. 5.11)

$$\frac{1}{q_f} = \frac{-\frac{z_R^2}{f}\left(1 - \frac{z'}{f}\right) + \left(1 - \frac{z_0}{f}\right)\left(z' + z_0 - \frac{z' z_0}{f}\right) - i z_R \left\{\frac{1}{f}\left(z' + z_0 - \frac{z' z_0}{f}\right) + \left(1 - \frac{z_0}{f}\right)\left(1 - \frac{z'}{f}\right)\right\}}{z_R^2 \left(1 - \frac{z'}{f}\right)^2 + \left(z' + z_0 - \frac{z' z_0}{f}\right)^2}. \tag{5.47}$$

And for the beam waist position of the focused beam, the condition

$$z' = z_{0f}: \quad -\frac{z_R^2}{f}\left(1 - \frac{z_{0f}}{f}\right) + \left(1 - \frac{z_0}{f}\right)\left(z_{0f} + z_0 - \frac{z_{0f} z_0}{f}\right) = 0$$

follows. Solving this equation for $1/z_{0f}$ results in the *focusing equation for Gaussian beams*, which is the equivalent to the lens equation in ray optics:

$$\frac{1}{z_{0f}} = \frac{1}{f} - \frac{1}{z_0} \frac{1}{1 + \frac{z_R^2}{z_0 (z_0 - f)}}. \tag{5.48}$$

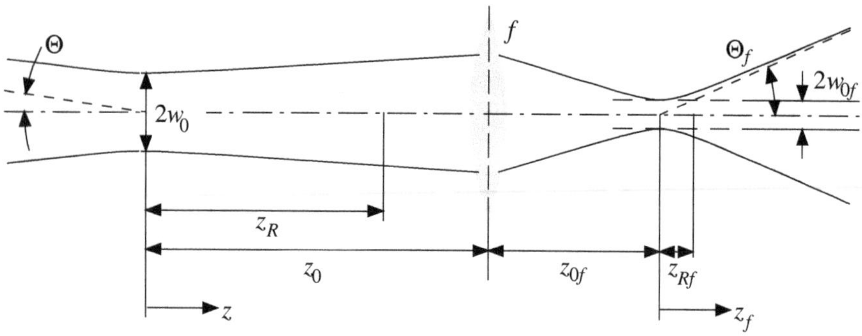

Fig. 5.11 Focusing a Gaussian beam by a thin lens with the focal length f. The parameters of the focused beam are distinguished by an additional index "f"

In this equation, z_0 is the object distance and z_{0f} the image distance. In comparison to the lens equation a corrective term has been added to the reciprocal of the object distance, so that the image distance now depends on *three* parameters:

- the focal length f of the lens,
- the object distance z_0, and
- the Rayleigh length z_R.

For $z_R \to 0$ Eq. 5.48 becomes the ray optics' lens equation. For $z_R \neq 0$ the image distance z_{0f} of the Gaussian beam is shifted by an offset Δf against the focal length of the lens, or the image distance according to ray optics:

$$\Delta f = z_{0f} - f = \frac{(z_0 - f)f^2}{(z_0 - f)^2 + z_R^2} \tag{5.49}$$

The larger the distance between beam waist and lens is compared to the Rayleigh length, the smaller is Δf. The shift reaches its maximum for

$$z_0 = z_R + f \Rightarrow \Delta f_{\max} = \frac{f^2}{2z_R} = \frac{\lambda}{2\pi}\left(\frac{f}{w_0}\right)^2. \tag{5.50}$$

Typically, $\Delta f_{\max}$ is much smaller than f and only needs to be considered if highest precision is required.

For the beam waist radius of the focused beam, or the focus radius, the following results with Eqs. 5.36 and 5.47

$$-\frac{\lambda}{\pi w_{0f}^2} = \Im\left(\frac{1}{q_f(z_{0f})}\right) = \frac{-z_R\left\{\frac{1}{f}\left(z_{0f} + z_0 - \frac{z_{0f} z_0}{f}\right) + \left(1 - \frac{z_0}{f}\right)\left(1 - \frac{z_{0f}}{f}\right)\right\}}{z_R^2\left(1 - \frac{z_{0f}}{f}\right)^2 + \left(z_{0f} + z_0 - \frac{z_{0f} z_0}{f}\right)^2},$$

and applying Eq. 5.48

$$w_{0f} = \frac{w_0 f}{\sqrt{z_R^2 + (z_0 - f)^2}} \tag{5.51}$$

$w_0 = \sqrt{\frac{\lambda z_R}{\pi}}$ Beam waist radius of the original beam (before the lens)

follow. For an increasing object distance the focus radius decreases. The ratio of the waist radii before and after the lens w_{0f}/w_0 depends not only on the focal length and the object distance, but also on the Rayleigh length of the unfocused beam. For $z_R \to 0$ Eq. 5.51 becomes the magnification equation known from ray optics (see Eq. 4.49).

If the object distance is large compared to the focal length of the lens, Eq. 5.51 can be approximated to

$$w_{0f} \approx \frac{w_0 f}{\sqrt{z_R^2 + z_0^2}} = \frac{w_0^2 f}{z_R w_0 \sqrt{1 + \frac{z_0^2}{z_R^2}}} = \frac{\lambda f}{\pi w_L} \tag{5.52}$$

$w_L = w(z_0)$ Beam radius on the lens.

From this, the commonly used approximate equation follows for the divergence angle of the focused Gaussian beam:

$$\Theta_f = \frac{w_{0f}}{z_{Rf}} = \frac{w_L}{f} \tag{5.53}$$

z_{Rf} Rayleigh length of the focused beam.

The transformation of higher-order modes obeys exactly the same rules as does the Gaussian beam. Thus, as for the Gaussian beam, the beam parameter product for any higher-order mode also remains unchanged. This means the beam parameter product and the beam quality of any beam that can be expressed as a superposition of Hermite-Gaussian or Laguerre-Gaussian modes remain unchanged under ABCD transformations.

5.5.3 *Adjustment of the Focus Radius*

For most applications it is necessary to adjust the beam radius in the processing area to a specific value in order to achieve the anticipated results (see Table 5.2). For this, two parameters are available:

- The *focal length* of the lens. Reducing the focal length of the lens results in a smaller focus radius (see Eq. 5.53). However, the focal length cannot typically be reduced arbitrarily, as, e.g., a minimum distance to a work piece has to be maintained. In material processing this may be required to avoid damaging the optics by splashes, sparks, and fumes from the processed material.
- The *object distance*. Increasing the object distance leads to a larger beam radius on the focusing lens and thus to a reduced focus radius (see Eq. 5.53, Fig. 5.12). In order to maintain a compact design of the optical system, a *telescope* can be employed to widen the beam. An increasing number of optical elements, however, increases the alignment effort and the transmission losses of the optical system.

Adjusting the focus radius by varying the distance between the focusing lens and the laser is a simple approach that does not require additional equipment. The necessary object distance z_0 results from the beam radius on the focusing lens needed to obtain the targeted focus radius:

$$w_L \equiv w(z_0) = w_0\sqrt{1+\frac{z_0^2}{z_R^2}} \Rightarrow z_0 = z_R\sqrt{\frac{w_L^2}{w_0^2}-1}. \tag{5.54}$$

By Using a *telescope* instead of a single lens, the beam is expanded first before it is focused by the focusing lens. Expanding the beam with a telescope has the

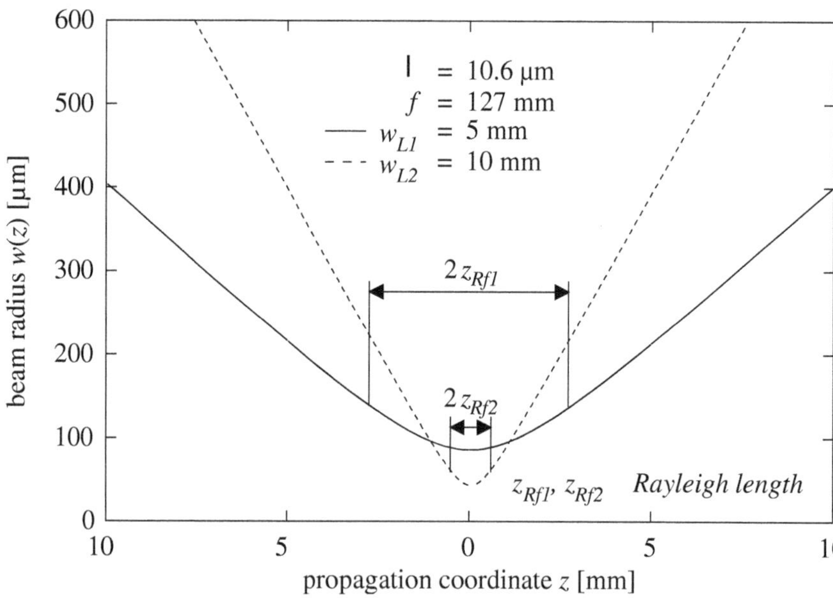

Fig. 5.12 The beam caustic of a Gaussian beam in the proximity of the focal plane, after focusing by a thin lens, for two different beam radii on the focusing lens

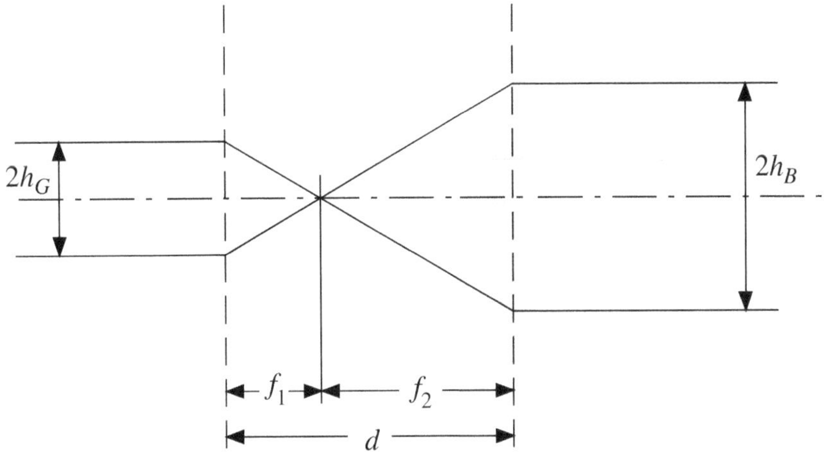

Fig. 5.13 Ray optics representation of the telescope. The ratio between the beam diameter on the image side h_B and the beam diameter on the object side h_G defines the expansion factor or lateral magnification of the telescope A_T

additional effect of increasing the Rayleigh length and thus reducing the divergence angle. This is particularly of advantage if the beam has to be guided over a longer distance from the beam source to the application area.

In its simplest configuration the telescope consists of two lenses. The focal length of a telescope is infinite; in the picture of ray optics, parallel incident rays are imaged on parallel rays again (Fig. 5.13). Since the focal length of the telescope is given by

$$\frac{1}{f} = \frac{1}{f_1} + \frac{1}{f_2} - \frac{d}{f_1 f_2} \tag{5.55}$$

f_1, f_2 Focal lengths of the respective lenses
d Distance of the lenses; length of the telescope

from $f \to \infty$ the condition

$$d = f_1 + f_2 \tag{5.56}$$

follows for the distance d between the two telescope lenses. Two parallel incident rays with a lateral distance h_G are imaged in parallel rays with distance h_B:

$$h_B = h_G \frac{f_2}{f_1} \Rightarrow A_T = \frac{h_B}{h_G} = \frac{f_2}{f_1}. \tag{5.57}$$

This defines the expansion factor or lateral magnification A_T of the telescope (Table 5.2).

Table 5.2 Adaption of the focus radius

System/measure	Characteristics/result
Nonoptimized system 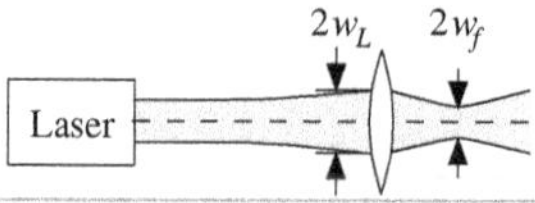	Small beam radius on the lens Large focal length $\Longrightarrow$ large focus radius
Reduced focal length 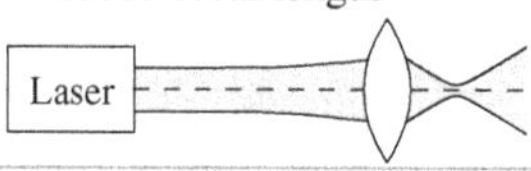	Small beam radius on the lens Short focal length $\Longrightarrow$ small focus radius
Increased object distance 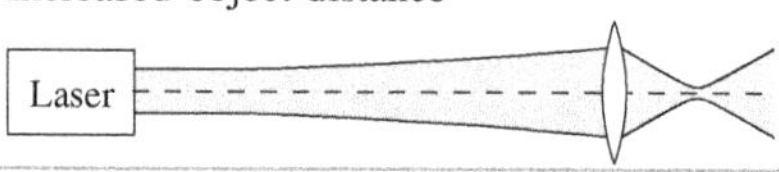	Large beam radius on the lens $\Longrightarrow$ small focus radius
Telescope 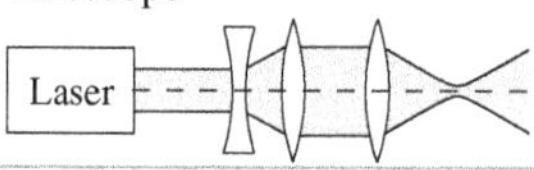	Short object distance Large beam radius on the focus lens $\Longrightarrow$ small focus radius

A reduced focal length as well as an increased object distance lead to a reduced focus radius, but with the disadvantage of a reduced working distance or an increased dimension of the optical system, respectively. By using a telescope both the disadvantages can be avoided

The focal length of the telescope can be set to a finite value by adjusting the distance between the two telescope lenses:

$$d = f_1 + f_2 + \Delta d \Rightarrow f = -\frac{f_1 f_2}{\Delta d}. \tag{5.58}$$

The transformation of a Gaussian beam by a telescope can be described using the complex beam parameter (Eq. 5.36) and the ABCD law (Eq. 5.38):

$$q = \frac{A_T q_0 + B_T}{C_T q_0 + D_T}. \tag{5.59}$$

The transfer matrix of the telescope $\mathbf{M}_T$ is composed of the matrices for the first lens, the propagation by the distance d between the lenses, the second lens and the propagation by a distance L behind the telescope:

$$\mathbf{M}_T = \begin{bmatrix} A_T & B_T \\ C_T & D_T \end{bmatrix} = \mathbf{M}_P(L) \cdot \mathbf{M}_f(f_2) \cdot \mathbf{M}_P(d) \cdot \mathbf{M}_f(f_2). \tag{5.60}$$

Depending on the application, L can, e.g., be the distance to the focus lens focusing the beam on the work piece. Using the matrices $\mathbf{M}_P$ and $\mathbf{M}_f$ as given in Eqs. 5.39 and 5.42 and approximating to the first order in Δd ($d = f_1 + f_1 + \Delta d$ and assuming $\Delta d \ll d$) result in the following parameters of the transformed beam behind the telescope:

$$\begin{aligned} w_{0T} &= w_0 \left|\frac{f_2}{f_1}\right| \cdot \left\{1 + \frac{\Delta d}{f_1}\left(1 - \frac{z_0}{f_1}\right)\right\}, \\ z_{RT} &= \frac{\pi w_{0T}^2}{\lambda} = z_R \left\{\frac{f_2}{f_1}\left[1 + \frac{\Delta d}{f_1}\left(1 - \frac{z_0}{f_1}\right)\right]\right\}^2, \\ z_{0T} &= \frac{(f_1 + f_2)f_2}{f_1} - z_0\frac{f_2^2}{f_1^2} + \Delta d\frac{f_2^2}{f_1^4}\left\{(z_0 - f_1)^2 - z_R^2\right\}. \end{aligned} \tag{5.61}$$

w_{0T} Beam waist radius of the transformed beam

z_{RT} Rayleigh length of the transformed beam

z_{0T} Distance from the telescope's exit plane to the beam waist of the transformed beam.

For $\Delta d = 0$, thus the telescope with infinite focal length, these expressions reduce to

$$\begin{aligned} w_{0T} &= w_0\left|\frac{f_2}{f_1}\right| = w_0|A_T|, \\ z_{RT} &= z_R\frac{f_2^2}{f_1^2} = z_R A_T^2, \\ z_{0T} &= \frac{(f_1 + f_2)f_2}{f_1} - z_0\frac{f_2^2}{f_1^2} = (f_1 + f_2)A_T - z_0 A_T^2. \end{aligned} \tag{5.62}$$

Figure 5.14 shows the expansion of a Gaussian beam by a telescope and its subsequent focusing by an additional lens. In this case, the first lens of the telescope is a diverging lens, i.e., $f_1 < 0$. This reduces the total length of the telescope.

When a telescope with very large magnification is used, a beam with a very large Rayleigh length and very low divergence can be generated. This allows the focusing lens to be moved within a wide range without a noticeable change of the beam radius on the lens, thus resulting in a constant focus radius. In this way, e.g., in laser cutting machines with so-called "flying optics," work pieces with large dimensions can be processed.

5.5.4 *Influence of Spherical Aberrations*

In practice, the beam expansion on the objective and, thus, the reduction of the focus radius is limited by the increasing influence of lens aberrations. Especially in this context, *spherical aberrations* play a role.

In Sect. 4.7.1, for the spherical aberration the term

$$\Delta r = B w_L^3 \tag{5.63}$$

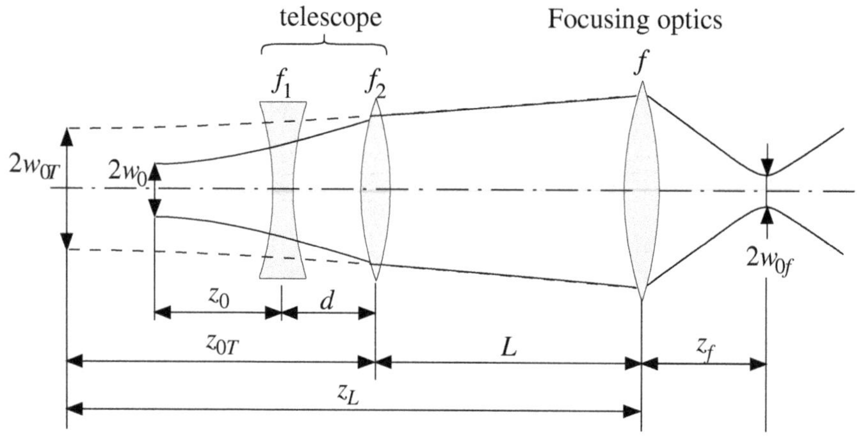

Fig. 5.14 Expansion of a Gaussian beam by a telescope and subsequent focusing by a lens

B Seidel coefficient

has been derived (cf. Eq. 4.60). The Seidel coefficient depends on the focal length, the material, and the shape of the lens:

$$B = \frac{8K_L}{f^2}. \tag{5.64}$$

K_L Lens parameter depending on refractive index and lens shape.

When spherical aberrations are considered and the approximation according to Eq. 5.52 is used, the focus radius is given by

$$w_{0f} = \frac{\lambda f}{\pi} \frac{1}{w_L} + \frac{8K_L}{f^2} w_L^3. \tag{5.65}$$

Typically, the lens parameter K_L is in the range between 0.01 and 0.2. Table 5.3 gives sample values of K_L for lens materials and shapes commonly used for the focusing of CO_2 laser beams.

From Eq. 5.65 it can be seen that—beyond a certain value for the beam radius on the lens, which depends on the focal length f and the lens parameter K_L—the focus radius does not continue to decrease, but rather increases again (see Fig. 5.15). The minimum focus radius is achieved if the beam radius on the lens assumes the optimum value $w_{L,\text{opt}}$:

$$\frac{\mathrm{d}}{\mathrm{d}w_L} w_{0f} = 0 \Rightarrow w_{L,\text{opt}} = \sqrt[4]{\frac{\lambda f^3}{24\pi K_L}}. \tag{5.66}$$

Table 5.3 K_L values of different lenses for the focusing of CO_2 laser beams

Lens material	Refractive index n	Lens shape	K_L
KCl	1.46	Planar convex	0.2350
ZnSe	2.40	Meniscus	0.0312
GaAs	3.27	Meniscus	0.0139

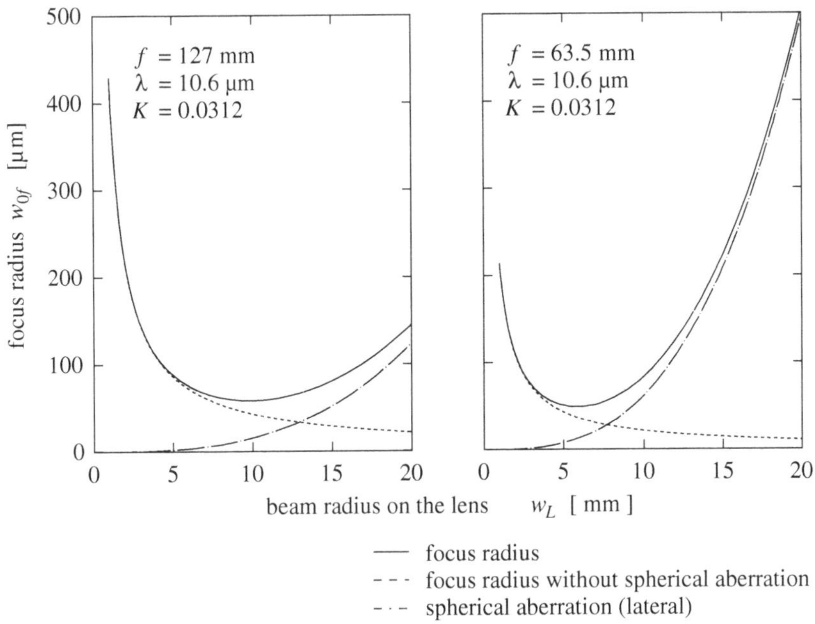

Fig. 5.15 Focus radius of a Gaussian beam after focusing by a thin Zinc Selenide lens, as a function of the beam radius on the lens, for two different focus lengths. Spherical aberrations lead to an increase of the focus radius if the beam radius on the lens is too large

The minimum possible focus radius is obtained by substituting $w_{L,\mathrm{opt}}$ into Eq. 5.65.

If, on the other hand, the lens aberrations are neglected, the minimum focus radius is achieved with the smallest possible ratio between f and w_L. Since in practice the focal length of the lens cannot be significantly smaller than the lens diameter, in this case the minimum achievable focus radius is in the order of the wavelength:

$$f \approx 2w_L \Rightarrow w_{0f} \approx \frac{2}{\pi}\lambda \approx \lambda. \tag{5.67}$$

Typically, the focus radii obtained with spherical lenses are significantly larger. Lens aberrations, however, can be reduced using corrected multilens focusing optics or lenses with aspheric profiles.

Chapter 6
Optical Resonators

Next to the gain medium, the optical resonator represents the second essential component of a laser. In general, a resonator is understood as a system able to oscillate that reacts to specific, discrete frequencies, the *eigen* or *resonance frequencies*, with maximum oscillation amplitude at sinusoidal excitation. Resonance phenomena have been observed and used in the most varied sectors of physics and technology. Music instruments, for example, are mechanical resonance systems: the sound is produced by exciting mechanical resonance vibrations and their transmission to the atmosphere. Microwave and laser resonators, however, represent electromagnetic resonance systems. In this case, the resonance vibrations are oscillations of the electromagnetic field.

Independent of the nature of resonance systems, the resonance frequencies are fundamentally determined by the systems' geometric dimensions. Common to conventional resonators (music instruments, microwave resonators, etc.) are their dimensions, which rest in the magnitude of wavelengths. In contrast to this, the dimensions of laser resonators are very large as compared to the wavelength, which means that a very high harmonic or *order* of the fundamental mode is excited in the laser resonator.

The tasks of the laser resonator are

- to feed the radiation emitted from the active medium back into the medium for further amplification, and
- to select one or several modes from the many possible self-oscillations or *modes* of the radiation field, which can be distinguished by *frequency* and *propagation direction*.

The selection of the eigenfrequencies occurs by the feedback and superpositioning of the backcoupled partial waves. The eigenfrequencies are determined by the constructive interference of the corresponding partial waves, while all the other frequencies lead to destructive interference. The decay time of the individual partial waves in the resonator determines the *frequency width* of the self-oscillations and thus the *quality* of the resonator.

R. Poprawe et al., *Tailored Light 1*, RWTHedition,
DOI 10.1007/978-3-642-01234-1_6

The laser resonator is setup as an *open resonator*[1] in order to select a preferable propagation direction. Open optical resonators consist of two mirrors, whose distance from each other is large compared to their diameters. This way, only such modes are fed back and amplified whose propagation direction only minimally deviates from the direction of the optical axis of the two-mirror system. To outcouple the laser beam, one of the mirrors is partially transparent.

Since the optical resonator determines the spectral distribution as well as the angular distribution of the laser radiation, it can be seen as an element of the laser which determines its quality.

6.1 Eigenmodes of the Electromagnetic Field

6.1.1 *Eigenmode of a One-Dimensional Resonator*

Completely confined by two completely reflecting mirrors, a one-dimensional area of the length L should initially serve as a basic model for the confinement of light waves in a resonator. In the interior of this area, the one-dimensional wave equation in a vacuum is valid, which means that the general solutions are plane waves

$$E = \left(A\mathrm{e}^{ikz} + B\mathrm{e}^{-ikz}\right)\mathrm{e}^{-i\omega t}, \quad k = \frac{\omega}{c_0}. \tag{6.1}$$

c_0 Speed of light in vacuum

At the boundary surfaces, the electrical field has to fulfill specific boundary conditions. For metallic walls which are assumed to be ideal conductors, the electric field vanishes there

$$E(z=0) = E(z=L) = 0. \tag{6.2}$$

If the ansatz of plane waves is used here, the following conditions result:

$$\begin{aligned} A + B &= 0 \\ A\mathrm{e}^{ikL} + B\mathrm{e}^{-ikL} &= 0. \end{aligned} \tag{6.3}$$

This system of linear equations is only solvable if

$$\mathrm{e}^{2ikL} = 1 \quad \Rightarrow \quad kL = n\pi, \quad n = 0, 1, 2, \ldots \tag{6.4}$$

[1]There are also closed optical resonators; they do not, however, lead to a mode selection: a closed resonator is nothing other than a cavity radiator and emits *thermal radiation* if it is in thermal equilibrium with its environment.

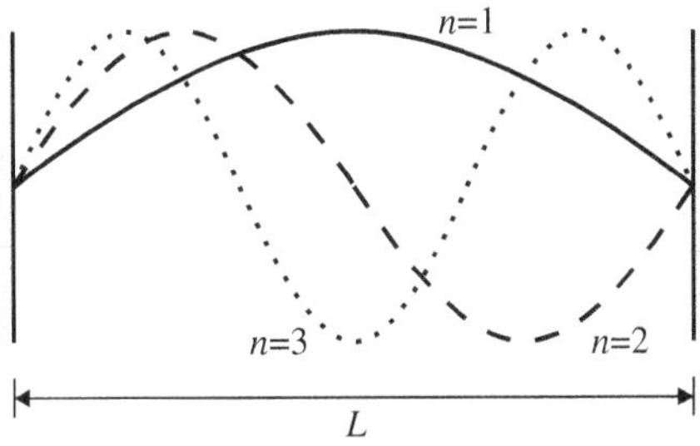

Fig. 6.1 The three lowest-order eigenmodes of a resonator of length L

is fulfilled. The wave numbers follow from the condition for solvability (eq. 6.4). The solution of the equation system, in contrast, delivers the coefficients A and B. Since the equation $A = -B$ is valid here, the following *eigenmodes* result:

$$E_n = E_{0,n} \sin(k_n z) \mathrm{e}^{-i\omega_n t}, \quad k_n = n\frac{\pi}{L}, \quad \omega_n = c_0 k_n, \quad n = 1, 2, 3, \ldots \tag{6.5}$$

These are standing waves. Due to the boundary conditions, the solution E = const. is impossible; the lowest eigenfrequency is $\omega_1 = c_0\,\pi/L$. The distance of neighboring modes in the wave number or frequency space amounts to

$$\Delta k = \frac{\pi}{L} \text{ bzw}, \quad \Delta\omega = c_0 \frac{\pi}{L}, \quad \Delta\nu = \frac{\Delta\omega}{2\pi} = \frac{c_0}{2L}. \tag{6.6}$$

Hence, the number of eigenmodes in a frequency interval $\mathrm{d}\nu$ amounts to (Fig. 6.1)

$$\mathrm{d}n = \frac{\mathrm{d}\nu}{\Delta\nu} = \frac{2L}{c_0}\mathrm{d}\nu = 2\frac{L}{\lambda}\frac{\mathrm{d}\nu}{\nu}. \tag{6.7}$$

6.1.2 Eigenmodes of a Rectangular Cavity

Now, a three-dimensional, rectangular cavity with edge lengths L_x, L_y, and L_z can be discussed. The three-dimensional wave equation in free space is valid in the interior. Through a separation ansatz

$$E(x, y, z; t) = E_0 f(x) g(y) h(z) \mathrm{e}^{-i\omega t} \tag{6.8}$$

the three-dimensional problem can, however, be reduced to the one-dimensional, previously solved case. By using the results of the previous section, the *eigenmodes of the three-dimensional, rectangular cavity* follow:

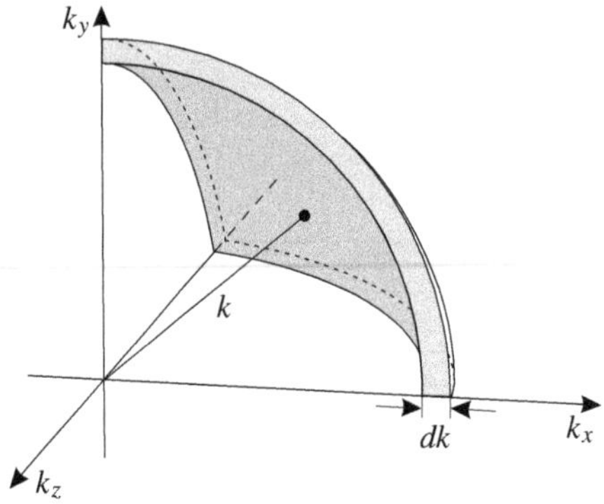

Fig. 6.2 The spherical shell in k-space: k is the radius of the spherical shell, dk its thickness

$$E_{lmn} = E_{0,lmn} \sin\left(k_{x,l}x\right) \sin\left(k_{y,m}y\right) \sin\left(k_{z,n}z\right) \mathrm{e}^{-i\omega_{lmn}t}$$
$$\text{mit} \quad \vec{k}_{lmn} = \begin{pmatrix} k_x \\ k_y \\ k_z \end{pmatrix} = \begin{pmatrix} l\pi/L_x \\ m\pi/L_y \\ n\pi/L_z \end{pmatrix}, \quad \omega_{lmn} = c_0 k_{lmn}, l, m, n = 1, 2, 3, \ldots \tag{6.9}$$

The number of modes dn in a frequency interval dν is now determined by the division of the volume of the respective spherical shell in k-space by the specific volume of a mode. In addition, a factor of two must be added to account for the two possible polarization directions of the wave (Fig. 6.2).

The volume of a mode in k-space results from the multiplication of the distances of the modes in all three spatial directions[2]:

$$V_{k,0} = \frac{\pi}{L_x} \frac{\pi}{L_y} \frac{\pi}{L_z} = \frac{\pi^3}{V}. \tag{6.10}$$

The volume of the spherical shell in k-space amounts to

$$\mathrm{d}V_k = 4\pi k^2 \mathrm{d}k \quad \text{mit} \quad k = \frac{2\pi\nu}{c_0}, \quad \mathrm{d}k = \frac{2\pi \mathrm{d}\nu}{c_0}. \tag{6.11}$$

Now, however, what must be considered is that for the eigenmodes of Eq. 6.9, only positive indices, l, m, n and thus only the k vectors with positive components are counted since the solutions with negative wavevector components do not represent linearly independent solutions. Only the segment of the complete spherical shell in the first octant, thus an eighth of the volume, has to be considered. For the *mode number* dn in the frequency interval dν, it follows that:

---

[2]Often the expression $(2\pi)^3/V$ is used as the volume of a mode. The additional factor 2^3 results from the use of periodic boundary conditions resulting in travelling waves instead of the boundary conditions used here which result in standing waves. Nothing changes to the mode number, however, since positive and negative values for k are allowed then for all spatial directions which compensate for the additional factor, cf. below.

$$dn = 2\frac{dV_k}{8V_{k,0}} = 2V\frac{4\pi k^2 dk}{8\pi^3} = 8\pi V\frac{\nu^2 d\nu}{c_0^3} = 8\pi\frac{V}{\lambda^3}\frac{d\nu}{\nu}. \tag{6.12}$$

For $dn = 1$, the distance of neighboring modes can be derived from this

$$\Delta\nu = \frac{1}{8\pi V}\frac{c_0^3}{\nu^2}. \tag{6.13}$$

Often the *density of states* is used in place of Eq. 6.12. The density of states indicates the number of eigenmodes per frequency interval and unit volume

$$D(\nu) = \frac{1}{V}\frac{dn}{d\nu} = \frac{8\pi\nu^2}{c_0^3}. \tag{6.14}$$

In many branches of physics, one often falls back upon counting the modes of a volume according to Eq. 6.12, for example, in Planck's law of black-body radiation or in quantum physics.

6.2 Selection of Modes and Resonator Quality

One of the two essential tasks of the resonator is mode selection. Since the laser should only amplify and emit light of one frequency and in the fundamental mode as far as possible—hence as a Gaussian beam—only a small amount of eigenmodes, from the many in the resonator, may oscillate and the others must be suppressed.

In this context, a distinction is often made between *longitudinal* and *transverse* modes. If the z-axis corresponds to the emission direction of the laser, then the part of the eigenmode dependent upon z is referred to as the longitudinal mode, the parts dependent on x and y as transverse modes. For this it is assumed that the separation of variables (z-coordinate and transverse coordinates) is approximately possible for the eigenmodes of the resonator, which is not always the case when viewed strictly.

The selection of a special longitudinal mode is referred to as *frequency selection*, since the emitted frequency is established through the choice of this mode. The transverse modes, in contrast, indicate the deviation of the k-vector from the emission direction. A high transversal mode order means that k-vectors with large transverse components (plane waves with large tilts with respect to the optical axis) contribute to the field. The selection of specific transverse modes is referred to as *directional selection*; only those waves with a specific propagation direction are chosen (Fig. 6.3).

As a rule, the longitudinal and transverse parts of modes are not completely independent of each other: different transverse modes in general exhibit different frequencies. The concepts of frequency and directional selection should be understood only as a rough organizing principle, since frequency selection also occurs as a side effect of directional selection, albeit in a smaller order of magnitude.

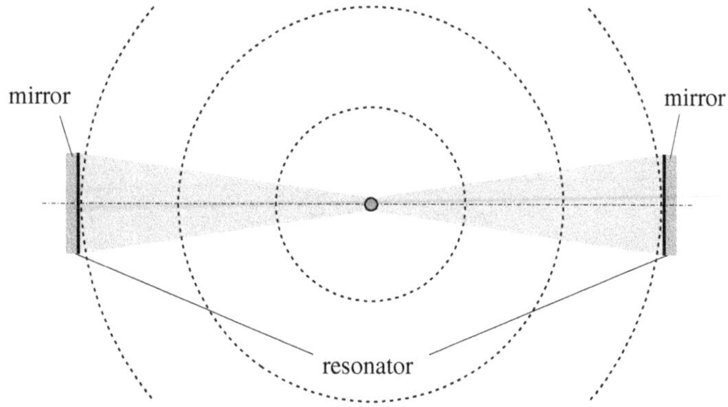

Fig. 6.3 Selection of direction. From the waves emitted in all directions, only a part is caught by the resonator mirrors, which lies in a specific solid angle

For the emitted beam, the transversal modes determine the mode order of the beam, hence the decomposition of the beam into Hermite-Gaussian or Laguerre-Gaussian modes.

6.2.1 The Open Resonator

At a wave length of 10 μm, a typical resonator volume of 10^{-3} m^3 and an equally common frequency width of 1 GHz, a number of nearly 8.5×10^8 enclosed eigenmodes follow from Eq. 6.12. From this enormous number of modes, only few as possible are selected by means of an appropriate construction of the resonator.

The comparison of Eq. 6.12 with Eq. 6.7 shows that the number of eigenmodes is lowered by the factor

$$\frac{\mathrm{d}n^{(1\mathrm{d})}}{\mathrm{d}n^{(3\mathrm{d})}} \approx \frac{\lambda}{L_x}\frac{\lambda}{L_y} \tag{6.15}$$

when one spatial dimension is considered instead of three. Because of the macroscopic expansion of typical laser resonators, this factor is very small: if the resonator taken from the above example is assumed to be cubic, then this factor amounts to 10^{-8}. One can see from this that by reducing the resonator to one dimension, the number of modes could already be drastically decreased.

Practically, the one-dimensional resonator can only be approached by removing the transversal resonator walls, or by assuring that these do not reflect. Such a resonator, which then only consists of mirrors on the head ends, is called an *open resonator*. Sooner or later, beams that are not axially parallel encounter one of the open sides and thus leave the resonator. This procedure is referred to as *directional selection*.

6.2.2 Frequency Selection: The Fabry-Perot-Resonator

The simplest resonator is the *Fabry-Perot resonator*. It consists of two plane mirrors, arranged parallel to each other, whose dimensions are so large that diffraction phenomena can be neglected. At least one of the mirrors is semi-reflective, in order to enable a outcoupling of the laser radiation. The longitudinal eigenmodes of this resonator are a good approximation of those of the one-dimensional area of Sect. 6.1.1[3];

$$E_n = E_{0,n}\sin(k_n z)\mathrm{e}^{-i\omega_n t}, \quad k_n = n\frac{\pi}{L}, \quad \omega_n = c_0 k_n, \quad n = 1,2,3,\ldots \tag{6.16}$$

To investigate the mechanism of frequency selection in the resonator in more detail, the transmission of the Fabry-Pérot resonator for a signal coupled into the resonator through one of the mirrors signal is initially determined.

For this, the total wave arising from the multiple reflection at mirrors is determined. The mirrors are characterized by the amplitude reflection coefficients r_1 and r_2, or by the amplitude transmission coefficients t_1 and t_2. For the sake of simplicity, these coefficients are assumed to be real-valued. The laser medium is taken into account by a simple amplification factor to estimate the influence the amplification has upon the frequency selectivity

$$V = \frac{E(L)}{E(0)}. \tag{6.17}$$

V *Amplification factor*

The amplification factor indicates the increase or decrease in the field intensity after passing through the resonator. For $V < 1$, the absorption predominates in the medium, for $V > 1$ amplification occurs.

Now an input signal E_0 is coupled into the resonator from one side. In Fig. 6.4 the development of the electric field is depicted according to each reflection at one of the mirrors. One can derive from this that the total transmitted field is given through

$$E_T = t \cdot E_0, \quad t = t_1 t_2 V \mathrm{e}^{ikL} \sum_{m=0}^{\infty} \left(r_1 r_2 V^2\right)^m \mathrm{e}^{2mikL}. \tag{6.18}$$

[3] The incomplete reflection on the mirrors leads to the electric field no longer completely vanishing at the mirror positions. The boundary conditions of the area have to be modified correspondingly. This leads to a small shift of the eigenfrequencies.

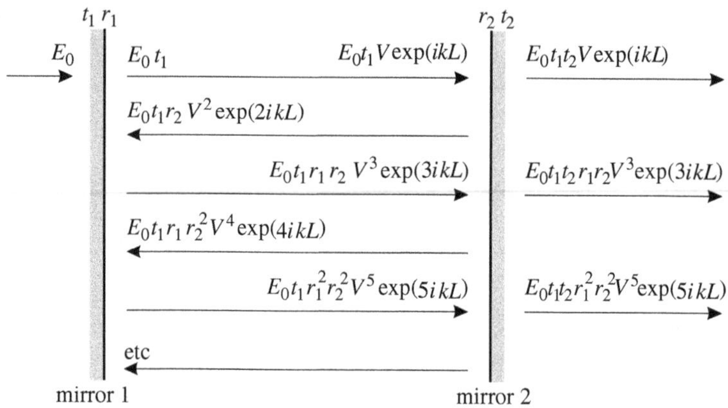

Fig. 6.4 Schematics of the multiple reflection in the Fabry–Pérot resonator

The amplitude transmission coefficient t of the system is complex in general. Using the summation formula for the geometric series,

$$\sum_{m=0}^{\infty} x^m = \tfrac{1}{1-x}, \quad |x| \leq 1\,, \tag{6.19}$$

t becomes to

$$t = \frac{t_1 t_2 V e^{ikL}}{1 - r_1 r_2 V^2 e^{2ikL}}. \tag{6.20}$$

Now two aspects of this expression can be discussed.

6.2.3 Eigen Modes and the Threshold of Self-Excitation

Initially the transmission becomes infinite when the denominator in Eq. 6.20 disappears. Then, the output signal E_t is independent of the input signal E_0. This designates the *threshold for self-excitation* or laser operation

$$1 - r_1 r_2 V^2 e^{2ikL} = 0. \tag{6.21}$$

For the phase the condition for the eigenmode of the resonator results from this

$$2kL = n \cdot 2\pi \quad \Rightarrow \quad k_n = n \cdot \tfrac{\pi}{L}, \tag{6.22}$$

which yields the modes known already from Sect. 6.1.1.[4] From this value, the *threshold conditions* for the amplification result in

$$r_1 r_2 V^2 = 1, \tag{6.23}$$

which means that the amplification has to compensate for the reflection losses of the mirrors, hence the outcoupling. If a simple, exponential growth is assumed for the amplification, what results is

$$V = e^{(g-\alpha)L} \quad \Rightarrow \quad g = \alpha - \tfrac{1}{2L}\ln(r_1 r_2)\,. \tag{6.24}$$

g Gain coefficient
α Absorbtion coefficient

In 1958, Schawlow and Townes formulated this threshold condition for laser operation. The gain coefficient g and the absorption coefficient a will be discussed in the following chapters.

The region above the threshold condition cannot be described in this context, since the self-exciting system oscillates independently of the input signal, and the concept of transmission cannot, therefore, be used. Mathematically, this is expressed in the fact that the geometrical series expansion used in Eq. 6.19 only converges as long as $[r_1 r_2 V^2] \le 1$ is fulfilled.

6.2.4 Line Width and Resonator Quality

The width of the transmission bands and the quality of the resonator is a second aspect that can be discussed on the basis of the transmission coefficient from Eq. 6.20. Since the width is determined using the intensity and not the field amplitude, one must calculate the *intensity transmission coefficient* beforehand

$$T = |t|^2 = \frac{(t_1 t_2 V^2)^2}{(1 - r_1 r_2 V^2)^2 + 4 r_1 r_2 V^2 \sin^2(kL)}\,. \tag{6.25}$$

For the maximums of the intensity transmission coefficient, the following is valid:

$$\sin kL = 0 \quad \Rightarrow \quad k_n = \tfrac{n\pi}{L}, \quad T_{\max} = \left(\tfrac{t_1 t_2 V^2}{1 - r_1 r_2 V^2}\right)^2. \tag{6.26}$$

[4]Nonreal reflection coefficients would lead to an additional, constant phase term in Eq. 6.22 and thus to a displacement of the eigen-modes.

Hence, they lie in a distance of

$$\Delta k = k_{n+1} - k_n = \frac{\pi}{L} \tag{6.27}$$

to each other, or rather, expressed in frequencies,

$$\nu_n = \tfrac{ck_n}{2\pi} = n\tfrac{c}{2L} \quad \Rightarrow \quad \Delta\nu = \tfrac{c}{2L}. \tag{6.28}$$

To obtain the full width at half maximum of δk or $\delta\nu$, the points where T has dropped to the half of its maximum are sought

$$\begin{aligned} T\left(kL = \tfrac{1}{2}\delta kL\right) &= \frac{1}{2}T_{\max} \Rightarrow 2\left(1 - r_1 r_2 V^2\right)^2 = \left(1 - r_1 r_2 V^2\right)^2 + 4r_1 r_2 V^2 \sin^2\left(\tfrac{1}{2}\delta kL\right) \\ &\Leftrightarrow \sin\frac{1}{2}\delta kL = \frac{1 - r_1 r_2 V^2}{2\sqrt{r_1 r_2}V} \\ &\Leftrightarrow \sin\left(\pi\frac{L}{c}\delta\nu\right) = \frac{1 - r_1 r_2 V^2}{2\sqrt{r_1 r_2 V^2}}. \end{aligned} \tag{6.29}$$

Close to the laser threshold the relation

$$r_1 r_2 V^2 \approx 1 \Rightarrow 1 - r_1 r_2 V^2 << 1, \tag{6.30}$$

holds, which means that the expression on the right side of Eq. 6.29 is small, so that the sine can be approximated by its argument and the square root in the denominator by one. Accordingly the approximate expression for the *line-width* follows (Fig. 6.5)

$$\delta\nu \approx \frac{c}{2L}\frac{1}{\pi}\left(1 - r_1 r_2 V^2\right). \tag{6.31}$$

At the laser threshold, the line width approaches zero; there the maxima are increased to infinity and are, therefore, infinitely narrow: there they take the form of a δ function.[5]

The relation of line spacing to line width is the *finesse* F of the resonator

$$F = \frac{\Delta\nu}{\delta\nu} = \frac{\pi}{1 - r_1 r_2 V^2}. \tag{6.32}$$

[5] δ-function refers to the Dirac delta function, defined according to $\delta(x - x_0) = \begin{cases} \infty & \text{für } x = x_0 \\ 0 & \text{sonst} \end{cases}, \int\limits_{-\infty}^{\infty} \delta(x - x_0)\mathrm{d}x = 1.$

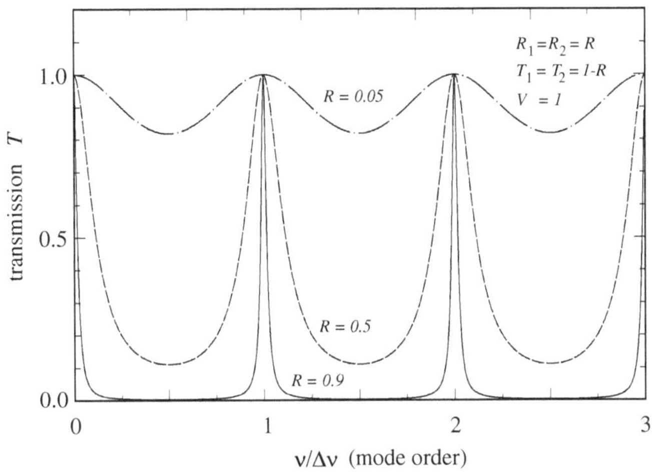

Fig. 6.5 Transmission of the Fabry–Perot resonator for various reflection coefficients of the mirror. Indicated here are the intensity reflection coefficient $R = |r|^2$

It approaches infinity close to the laser threshold according to Eq. 6.23; indeed, the linewidth of lasers is extraordinarily narrow.[6] The line width and the finesse of the empty resonator follow for an amplification factor of $V = 1$. A high finesse means a good frequency selection of the resonator.

At the same time, the line width of the resonator is a gauge for the radiation loss in the resonator per cycle. This is equivalent to mechanical oscillators for which the width of the resonance lines is also determined by the damping constants. The energy loss of the resonator per cycle is also expressed by the *resonator quality* Q:

$$Q = \Omega \frac{W}{\dot{W}} = \frac{c}{2L} \frac{W}{\dot{W}} \quad \left(\dot{W} \equiv \frac{\mathrm{d}W}{\mathrm{d}t} \right). \tag{6.33}$$

Q Resonator quality
Ω Resonator round-trip frequency
W Energy contained in the resonator

The energy W stored in the resonator is proportional to the mode frequency $\nu_n = \omega_n/2\pi$, and to the energy loss per cycle to the line width $\delta\nu$, so that the resonator quality corresponds to the reciprocal relative line-width

$$W \sim \nu_n, \frac{\dot{W}}{\Omega} \sim \delta\nu \Rightarrow Q_n = \frac{\nu_n}{\delta\nu} = \frac{n\pi}{1 - r_1 r_2 V^2}. \tag{6.34}$$

[6]The resulting line-width of the laser radiation is determined by the remaining spontaneous emission, which is not incorporated in Eq. 6.32.

n Mode order/arrangement
$V < 1$ Internal resonator losses

The resonator quality is normally specified for the empty resonator without any gain medium, which is why V only accounts for the losses in the resonator interior in this case, and therefore is always smaller than one. High reflection coefficients, or low losses in the resonator lead to a high quality.

6.3 Resonators with Spherical Mirrors

Calculating the eigensolutions of open optical resonators represents a mathematically difficult problem. Since the mirrors are spatially limited, diffraction occurs, and the propagation of the radiation field from mirror to mirror has to be described using the Kirschhoff diffraction integral. For the most part the calculations arising from this can only be solved numerically, particularly when the resonator mirrors are no longer planar and their curvature has be taken into account in the calculations.

The problem is simplified significantly when the diffraction at the mirrors are initially neglected. For this, it is assumed that the beam radius is considerably smaller than the radius of the mirrors; the validity of this assumption can be verified experimentally in individual cases. As a consequence of this assumption, it is consistent to assume highly directional beams and to use the SVE approximation. Great significance is, therefore, attached to the solutions of the wave equations in the SVE approximation derived in the last chapter, the Hermite-Gaussian and Laguerre-Gaussian modes, when the field distribution in resonators is described.

6.3.1 Beam Geometry in the Resonator

The propagation of Gaussian beams and their transformation through optical elements can be described using the ABCD law and the corresponding beam transfer matrices (cf. Sect. 5.4). To determine the eigenmodes of the resonator, a complete round-trip of the beam in the resonator has to be calculated. A complete round-trip consists of

1. The propagation by the resonator length L from mirror 1 to mirror 2,
2. The reflection at mirror 2,
3. The propagation from mirror 2 to mirror 1, and
4. The reflection at mirror 1 (Fig. 6.6).

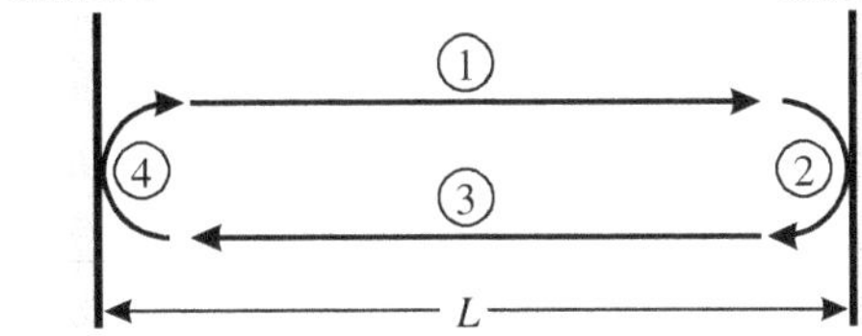

Fig. 6.6 Schematics of the resonator round-trip. *1* Propagation from mirror 1 to mirror 2; *2* reflection at mirror 2; *3* propagation back to mirror 1; *4* reflection at mirror 1

The beam transfer matrix for the propagation through a length z reads as follows:

$$\mathbf{M}_P(z) = \begin{bmatrix} 1 & z \\ 0 & 1 \end{bmatrix}, \tag{6.35}$$

and for the reflection at a spherical mirror with the radius of curvature R, it is

$$\mathbf{M}_S(R) = \begin{bmatrix} 1 & 0 \\ -\frac{2}{R} & 1 \end{bmatrix}. \tag{6.36}$$

For the full round-trip in a resonator with length L and mirrors with radii of curvature R_1 and R_2, the following transfer matrix results:

$$\begin{aligned} \mathbf{M}_R &= \mathbf{M}_S(R_1) \cdot \mathbf{M}_P(L) \cdot \mathbf{M}_S(R_2) \cdot \mathbf{M}_P(L) \\ &= \begin{bmatrix} 1 & 0 \\ -\frac{2}{R_1} & 1 \end{bmatrix} \cdot \begin{bmatrix} 1 & L \\ 0 & 1 \end{bmatrix} \cdot \begin{bmatrix} 1 & 0 \\ -\frac{2}{R_2} & 1 \end{bmatrix} \cdot \begin{bmatrix} 1 & L \\ 0 & 1 \end{bmatrix}. \end{aligned} \tag{6.37}$$

With the so-called g parameters

$$g_1 = 1 - \frac{L}{R_1} \quad \text{and} \quad g_2 = 1 - \frac{L}{R_2}, \tag{6.38}$$

the matrix reads as follows:

$$\mathbf{M}_R = \begin{bmatrix} 2g_2 - 1 & 2Lg_2 \\ \frac{2}{L}(2g_1g_2 - g_1 - g_2) & 4g_1g_2 - 2g_2 - 1 \end{bmatrix}. \tag{6.39}$$

For the complete beam parameter q after j round-trips

$$q_{j+1} = \frac{Aq_j + B}{Cq_j + D} \quad \text{mit} \quad \mathbf{M}_R = \begin{bmatrix} A & B \\ C & D \end{bmatrix}. \tag{6.40}$$

j Number of resonator round-trips

is valid. A eigensolution q_E of the resonator is present when the beam parameter remains constant from cycle to cycle, thus when q_E fulfills the condition

$$q_{j+1} = q_j = q_E \quad \Rightarrow \quad q_E = \frac{Aq_E + B}{Cq_E + D}. \tag{6.41}$$

Solving for q_E yields

$$\begin{aligned} q_E &= \frac{1}{2C}\left(A - D \pm \sqrt{(D-A)^2 + 4BC}\right) \\ &= \frac{1}{2C}\left(A - D \pm \sqrt{(D+A)^2 - 4DA + 4BC}\right), \end{aligned} \tag{6.42}$$

and when using the relation valid for all beam transfer matrices

$$\det \mathbf{M} = AD - BC = 1 \tag{6.43}$$

q_E finally thus becomes

$$\begin{aligned} q_E &= \frac{1}{2C}\left(A - D \pm \sqrt{(A+D)^2 - 4}\right) \\ &= \frac{1}{2C}\left(A - D \pm i\sqrt{4 - (A+D)^2}\right). \end{aligned} \tag{6.44}$$

With the help of the definition of the complex beam parameter

$$q = z + iz_R, \quad z_R = \frac{\pi w_0^2}{\lambda} \tag{6.45}$$

z_R Rayleigh length
z Coordinate in the direction of propagation, measured from the beam waist

the Rayleigh length and the position of the beam waist can be determined (Fig. 6.7). Since the Rayleigh length is positive by definition, the algebraic sign in Eq. 6.44 has to be chosen correspondingly.

$$\begin{aligned} z_R &= \frac{\sqrt{4 - (A+D)^2}}{2C} = L\frac{\sqrt{g_1 g_2(1 - g_1 g_2)}}{g_1 + g_2 - 2g_1 g_2}, \\ z \equiv z_1 &= \frac{A - D}{2C} = -\frac{Lg_2(1 - g_1)}{g_1 + g_2 - 2g_1 g_2}. \end{aligned} \tag{6.46}$$

The order of the ray transfer matrices in Eq. 6.37 was chosen in such a way that the round-trip ends after the reflection at mirror 1, which means q_E is the beam parameter of the beam running from mirror 1 to mirror 2. Therefore, z is the distance of the first mirror from the beam waist. The distance of the beam waist to the second mirror is then

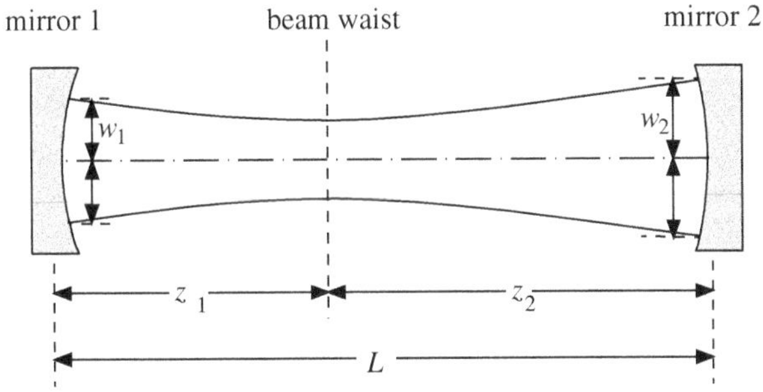

Fig. 6.7 The Gaussian beam in a spherical resonator

$$z_2 = z_1 + L = \frac{Lg_1(1-g_2)}{g_1+g_2-2g_1g_2}. \tag{6.47}$$

From the definition of the Rayleigh length, one obtains the beam waist of the beam

$$w_0 = \sqrt{\frac{z_R\lambda}{\pi}} = \sqrt{\frac{\lambda L}{\pi}} \cdot \frac{[g_1g_2(1-g_1g_2)]^{\frac{1}{4}}}{\sqrt{g_1+g_2-2g_1g_2}}. \tag{6.48}$$

The radii of curvature of the phase fronts at the mirrors are

$$R(z_{1,2}) = z_{1,2}\left(1+\frac{z_R^2}{z_{1,2}^2}\right) = z_R\left(\frac{z_{1,2}}{z_R}+\frac{z_R}{z_{1,2}}\right) = \frac{L}{1-g_{1,2}} = R_{1,2}, \tag{6.49}$$

and thus equal to the radii of curvature of the mirrors.

6.3.2 *The Stability Criterion*

One can read from Eq. 6.48 that a real and finite waist w_0 only results when the g parameters of the resonator fulfill the following condition

$$0 < g_1g_2 < 1. \tag{6.50}$$

This condition is referred to as the *stability criterion* for spherical resonators. Resonators whose g parameters fulfill the criterion are called *stable resonators*. In this case, the radiation field of the eigenfunctions of the resonator remains concentrated around the optical axis. If the g parameters do not fulfill the stability criterion, the resonator is thus *unstable*. For this, a distinction is made between a positive and negative branch, according to the algebraic sign of the product g_1g_2.

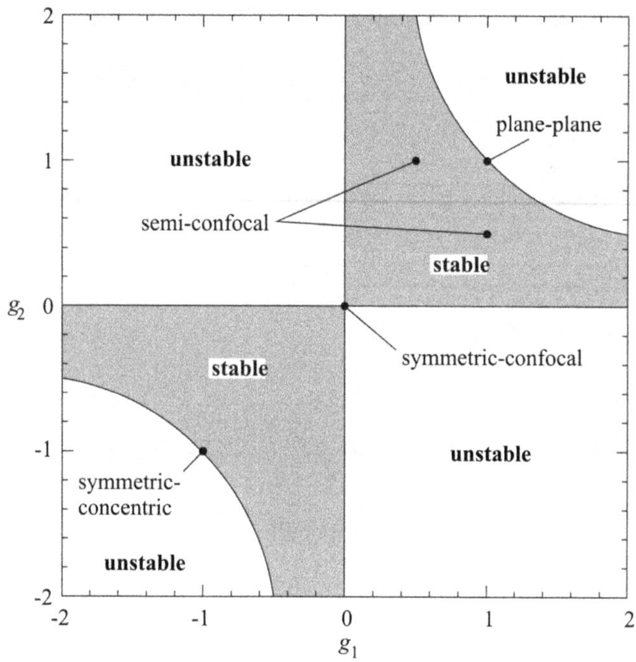

Fig. 6.8 The stability diagram for spherical resonators

The calculations made in the last section are only applicable for stable resonators, since only in this case can the beam radius remain small in comparison to the mirror dimensions and, therefore, can the diffraction be justifiably neglected. In the case of an unstable resonator, by contrast, the beam radius grows until it is limited by the fact that sufficiently enough radiation leaves the resonator because it passes one of the mirrors on the sides. This is then also used to outcouple the beam: the outcoupled laser beam is formed by radiation that passes the smaller mirror on the sides. With stable resonators, on the other hand, a semi-reflective mirror is used for the outcoupling.

In Fig. 6.8, the *stability diagram* for spherical resonators is illustrated. For this g_1 and g_2 are plotted against each other. Stable regions are found only in the first and third quadrants. Several pairs of g parameters, which correspond to special resonator configurations, are displayed.

Planer–planer resonator

Both mirrors are plane, which means a Fabry-Perot resonator is, in principle, present. What differs, however, is that the expansion of the mirror surfaces is so small that the diffraction has to be accounted for. Plane mirrors have an infinite radius of curvature R, which is why both g parameters for the planar–planar resonator are 1

$$R_{1,2} \to \infty \quad \Rightarrow \quad g_{1,2} = 1 - \frac{L}{R_{1,2}} = 1.$$

In this way the plane–plane resonator rests just outside the stability criterion; its eigenmodes are no longer definable with the procedure represented in Section.

A theoretical advantage of the plane–plane resonator is that the entire volume between the mirrors is covered by the beam, which means an optimal utilization of the active medium can be attained. For resonators with very large, but finite curvatures $R_{1,2} \gg L$, this advantage is mainly retained, yet at the same time these resonators are, however, stable. Such resonators are named *long-range resonators*.

Symmetric-confocal resonator
In the case of the symmetric-confocal resonator, both mirrors have a common focal point in the middle of the resonator. Since the focal length of a spherical mirror is $f = R/2$, then

$$f_{1,2} = \frac{L}{2} \quad \Rightarrow \quad R_{1,2} = L \quad \Rightarrow \quad g_{1,2} = 0.$$

This resonator also is located at the edge of the stable region (Fig. 6.9).

Symmetric-concentric resonator
The third symmetric resonator, which is located at the edge of the stability region, is the symmetric-concentric resonator. For this the radii of curvature of the mirrors are equal to half of the resonator length, which means the spherical surfaces of both

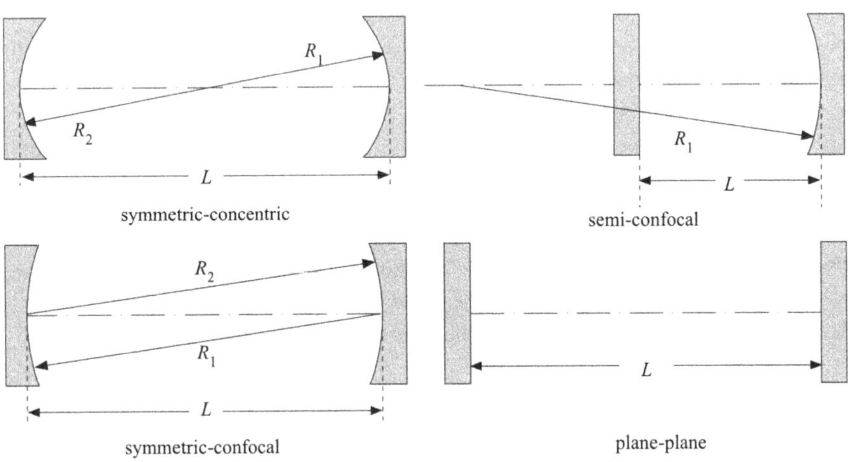

Fig. 6.9 Different configurations of optical resonators

mirrors have a common center in the middle of the resonator. The focal point of the mirrors are then at one-fourth or three-fourth of the resonator length.

The g parameters are

$$g_{1,2} = 1 - \frac{L}{\frac{L}{2}} = -1,$$

so that this resonator also lies at the edge of the stability region.

Resonators that are located exactly on the limit between stability and instability are especially sensitive to adjustment, since small deviations in the positions of the mirrors can already lead to the resonator becoming unstable and its emission behavior changing fundamentally.

Semi-confocal resonator

The semi-confocal resonator consists of a plane mirror and a spherical mirror, whereas the resonator length corresponds to the focal length of the mirror

$$R_1 = \infty, \quad f_2 = L \Leftrightarrow R_2 = 2L \quad \Rightarrow \quad g_1 = 1, \quad g_2 = \tfrac{1}{2}.$$

The semi-confocal resonator lies in the stable area. Its name results in it corresponding to a symmetric-confocal resonator halved by the plane mirror at the symmetry plane.

6.3.3 Eigenfrequencies of Stable Spherical Resonators

So far, the discussion of spherical resonators has only referred to the description by means of the complex beam parameter. This contains the information about the position of the beam waist and the Rayleigh length, having the same meaning as the divergence angle of the beam. This means that only the directional selection of the resonator has been discussed to date; the frequency or the wave-length were free parameters.

For the frequency selection of a spherical resonator, the same condition is basically valid as is for the rectangular cavity in Sect. 6.1.1. During a resonator round trip, the phase has to change by an integral multiple of 2π so that the waves are superimposed constructively in the resonator and form standing waves. For a Hermite-Gaussian mode TEM_{mn}, the phase on the optical axis is

$$\psi(z) = kz - (m + n + 1)\arctan\left(\frac{z}{z_R}\right), \tag{6.51}$$

so that the resonance condition reads as follows:

$$\psi(z_2) - \psi(z_1) = kL - (m+n+1)\left[\arctan\left(\frac{z_2}{z_R}\right) - \arctan\left(\frac{z_1}{z_R}\right)\right] = j\pi, \quad j = 0, 1, 2, \ldots. \tag{6.52}$$

$z_{1,2}$ Position of the first or second mirror
$L = z_2 - z_1$ Length of the resonator

From the relationships

$$\begin{aligned} \arctan x &= \arccos\left(\frac{1}{\sqrt{1+x^2}}\right), \\ \arccos x - \arccos y &= \arccos\left(xy - \sqrt{1-x^2}\sqrt{1-y^2}\right) \end{aligned} \tag{6.53}$$

and the frequency

$$\nu = \frac{c_0 k}{2\pi} \tag{6.54}$$

c_0 Speed of light in a vacuum

the *eigenfrequencies of the spherical resonator* result

$$\nu_{mn,j} = \frac{c_0}{2L}\left[j + (m+n+1)\frac{1}{\pi}\arccos\left(\sqrt{g_1 g_2}\right)\right]. \tag{6.55}$$

A corresponding expression can also be derived for the Laguerre-Gaussian modes. The radial and azimuthal orders p and l take the place of the mode orders n and m for the x and y direction

$$(m+n+1) \quad \rightarrow \quad (p+l+1). \tag{6.56}$$

One can see in Eq. 6.55 that the eigenfrequencies are not generally dependent on the longitudinal mode order j, but rather on the transversal orders m and n. The simultaneous oscillation buildup of several transversal modes can, therefore, lead to beats in the laser beam intensity. In Fig. 6.10, the frequency spectra of several spherical resonator types are depicted schematically.

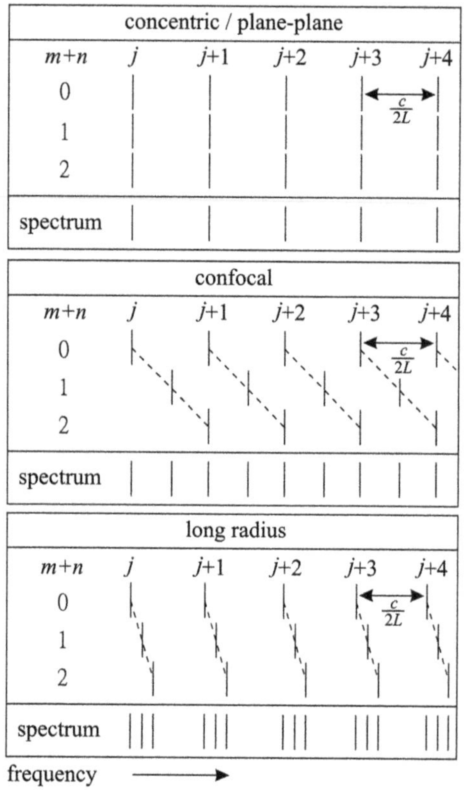

Fig. 6.10 Schematic representation of the frequency spectra of several spherical resonators

6.4 Influence of Mirror Boundaries

The eigenfunctions derived in the previous section represent good approximations only as long as the beam radius on the mirrors is significantly smaller than the mirror radius itself. Only in this case can the diffraction effects be neglected during the reflection at the mirror.

To more exactly determine the eigenmodes of open optical resonators, the diffraction problem has to be solved for these resonator types. In general, the field distribution second mirror can be calculated by inserting the field distribution on the first mirror in the Kirchhoff diffraction integral and propagating to the second mirror. As a rule, the radius of the mirror r_S is far smaller than the resonator length L, in order to improve the directional selection of the resonator. Principally, the paraxial approximation can, therefore, be applied for the beams in the resonator and the Kirchhoff diffraction integral can be simplified to the Fresnel integral.

6.4.1 The Diffraction Integral Between Curved Mirrors

The curved resonator mirrors complicate the procedure. The transformation of the field due to the mirror curvature can be represented similar to that of a phase shift imposed by a lens, but in this case it is simpler to modify the diffraction integral so that it directly images the curved surfaces onto each other.

For this one starts with the original Kirchhoff diffraction integral,

$$E(\vec{r}_2) = \frac{ik}{2\pi} \iint_{S_1} E(\vec{r}_1) \frac{\exp(ik\vec{R})}{R} \mathrm{d}A. \tag{6.57}$$

$E(\vec{r}_1)$, $E(\vec{r}_2)$ Electrical field on the first or second mirror, respectively
$\vec{R} = \vec{r}_2 - \vec{r}_1$ Difference vector from the test and integration points

The vector $\vec{r}_2$ lies on the surface of the second mirror and we integrated over the surface S_1 of the first mirror; R is the distance between $\vec{r}_2$ and the integration point:

$$R^2 = (\vec{r}_2 - \vec{r}_1)^2 = (x_2 - x_1)^2 + (y_2 - y_1)^2 + (z_2 - z_1)^2. \tag{6.58}$$

For the magnitude of $\vec{r}_2$, the following is valid:

$$r_2^2 = x_2^2 + y_2^2 + z_2^2 \quad \Rightarrow \quad z_2 = \sqrt{r_2^2 - x_2^2 - y_2^2} \approx r_2 - \frac{x_2^2 + y_2^2}{2r_2}. \tag{6.59}$$

If the zero point of the coordinate system is displaced so that it lies in the middle of the resonator, then this follows,

$$z_2 \to z_2 + r_2 - \frac{L}{2} \quad \Rightarrow \quad z_2 \approx \frac{L}{2} - \frac{x_2^2 + y_2^2}{2r_2} = \frac{L}{2} - \Delta z_2, \tag{6.60}$$

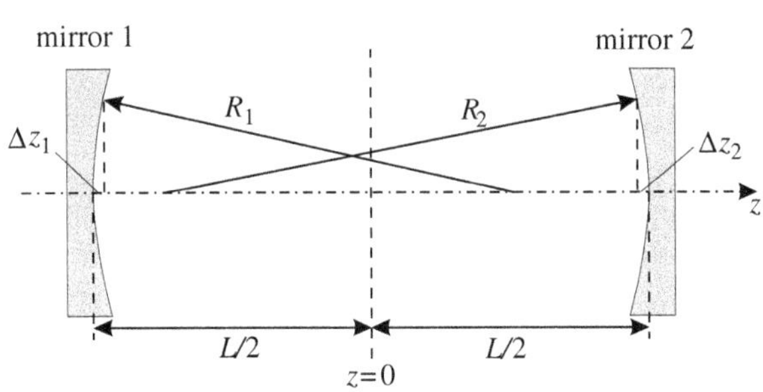

Fig. 6.11 Geometry to the diffraction integral between curved mirrors

and using the corresponding expression for z_1, the difference of the z coordinate can be written as (Fig. 6.11)

$$z_2 - z_1 = L - \frac{x_2^2 + y_2^2}{2r_2} - \frac{x_1^2 + y_1^2}{2r_1} = L - \Delta z_2 - \Delta z_1. \tag{6.61}$$

Squaring and neglecting terms of the fourth order in the x- and y-coordinates lead to

$$(z_2 - z_1)^2 \approx L^2 - L\left(\frac{x_1^2 + y_1^2}{r_1} + \frac{x_2^2 + y_2^2}{r_2}\right). \tag{6.62}$$

If this expression is inserted into Eq. 6.58 in turn, and if the g parameters defined in Eq. 6.38 are used, then

$$\begin{aligned} R^2 &= L^2 + g_1\left(x_1^2 + y_1^2\right) + g_2\left(x_2^2 + y_2^2\right) - 2x_1x_2 - 2y_1y_2 \\ \Rightarrow R &\approx L + \frac{g_1}{2L}\left(x_1^2 + y_1^2\right) + \frac{g_2}{2L}\left(x_2^2 + y_2^2\right) - \frac{1}{L}(x_1x_2 + y_1y_2). \end{aligned} \tag{6.63}$$

follows for the distance R in paraxial approximation.

The approximation is used for the phase term in the diffraction integral. In the denominator of the integrand, R is simply approximated using the resonator length L. From Eq. 6.57, the *diffraction integral between curved mirrors* in the Fresnel approximation reads

$$\begin{aligned} E(x_2, y_2, z_2) = &\frac{ik}{2\pi}\frac{e^{ikL}}{L} \iint_{S_1} E(x_1, y_1, z_1) \\ &\times \exp\left\{\frac{ik}{2L}\left[g_1\left(x_1^2 + y_1^2\right) + g_2\left(x_2^2 + y_2^2\right) - 2x_1x_2 - 2y_1y_2\right]\right\} dx_1 dy_1. \end{aligned} \tag{6.64}$$

As a prerequisite for the validity of this relation, the beam radius as well as the radii of curvature of the mirrors have to be small compared to the resonator length L.

6.4.2 *Eigenvalue Equation for Open Spherical Resonators*

In order to calculate the field after a complete round trip in the resonator, the same transformation must again be applied to the field distribution, just gained, on the second mirror. In the case of an eigenmode, the shape of the field distribution then remains unchanged and the field only changes by a constant factor Γ. The *eigenvalue equation for the open spherical resonator* reads as follows:

$$E(x_1, y_1, z_1) = \Gamma\left(\frac{ik}{2\pi}\frac{e^{ikL}}{L}\right)^2$$
$$\iint_{S_2}\iint_{S_1} K(x_1, y_1; x_2, y_2)K(x_2, y_2; x_1', y_1')E(x_1', y_1', z_1)dx_1'dy_1'dx_2dy_2,$$
$$K(x_1, y_1; x_2, y_2) = \exp\left\{\frac{ik}{2L}\left[g_1\left(x_1^2+y_1^2\right)+g_2\left(x_2^2+y_2^2\right)-2x_1x_2-2y_1y_2\right]\right\}. \tag{6.65}$$

Γ is the in general complex eigenvalue to the eigenfunction $E_1(x_1,y_1,z_1)$. The value of Γ is smaller or equal to one; it describes the resonator losses. Since the intensity and the energy of a wave are proportional to the square of the electric field, the losses are

$$\delta_B = 1 - |\Gamma|^2 \tag{6.66}$$

per round trip. These losses result form the fact that not the entire radiation reflected from a mirror falls upon the opposing mirror. A part of the radiation passes the mirror plane outside the mirror. The cause of this can be purely geometrical. For stable resonators, however, the diffraction is solely responsible for these losses. They are, therefore, called *diffraction losses*.

6.4.3 *Eigenmodes According to the Methods from* Fox *and* Li

Analytical solutions of the eigenvalue problem of Eq. 6.65 can only be determined for few special cases. In general, numerical methods have to be applied for these solutions. The most well-known method is the *method from* Fox *and* Li. For this method, an appropriate field distribution on a mirror is chosen and the transformation Eq. 6.65 is often applied to this, which means the distribution is subjected to numerous resonator round trips.

The original field distribution can be expanded in terms of the eigenmodes of the resonator,

$$E^0(x, y, z) = \sum_n a_n E_n(x, y, z). \tag{6.67}$$

For one eigenmode E_n

$$M \circ E_n = \Gamma_n E_n, \tag{6.68}$$

is valid, whereas M is the transformation according Eq. 6.65. This equation represents a single resonator round trip. For N round trips,

$$\underbrace{M \circ M \circ \ldots \circ M}_{n \text{ times}} \circ E_n \equiv M^N \circ E_n = \Gamma_n^N E_n, \tag{6.69}$$

results, such that the field distribution is given by

$$E^N(x, y, z) = \sum_n a_n \Gamma_n^N E_n(x, y, z) \tag{6.70}$$

after N round trips. The eigenmode with the largest eigenvalue, thus with the lowest losses, is damped the least. After a sufficient amount of resonator round trips, most of the modes are suppressed and only the mode with the lowest losses remains, the fundamental mode

$$N \to \infty: \quad E^N(x, y, z) \to E_0(x, y, z). \tag{6.71}$$

In practice, 100 cycles are typically necessary.

When the fundamental mode has been approximately determined in this way, it can be substracted from the initial distribution E^0, so that the initial distribution is only made up of the higher modes

$$E^0 \to E^0 - a_0 E_0 = \sum_{n=1}^{\infty} a_n E_n. \tag{6.72}$$

If the described procedure is now once again executed using the new initial distribution, only the eigenmode remains, which exhibits the second lowest losses.

In principle, this procedure can be continued for as many higher order modes as desired. Since, however, the errors of all the previously determined modes enter into the following step, the method quickly becomes less precise as the order grows. To determine the essential modes, the method from Fox and Li is, however, sufficient. Since it is relatively simple, this method belongs among the standard procedures to numerically determine modes.

6.5 Unstable Resonators

An important goal of laser development is to construct laser systems that offer both high output power as well as high beam quality and compact construction. The common technique to accomplish this consists in enlarging the cross section of the laser resonator, without, however, strongly impairing beam quality. With stable resonators the beam quality quickly falls when the resonator cross section rises—that is, when mirror radii increase—since modes of higher transversal orders begin to oscillate. The multiple folding of stable resonators represents a possible alternative. This has its limits, however, which are related to the sensitivity to adjustment and the poor utilization of the active medium, according to construction.

A further alternative is an *unstable resonator*, which enables better utilization of the active medium as well as a good beam quality. For unstable resonators,

$$g_1 g_2 \leq 0 \quad \text{or} \quad g_1 g_2 \geq 1. \tag{6.73}$$

is valid. In contrast to stable resonators, in which the beam distribution and thus also the beam radius reproduce themselves after each round trip, the beam radius of unstable resonators increases with each round trip. This leads to the beam radius growing beyond the mirror radius; the beam is outcoupled at the sides of the smaller mirror. The losses, or the outcoupled share of the radiation, are always very large with this procedure. Unstable resonators exhibit a series of problems which make their application in practice difficult

- In principle, the losses are high and typically lie from 30–50 % per round trip, which limits the use of unstable resonators on laser media with high gain (CO_2 or solid-state lasers).
- The sensitivity to adjustment is significantly higher than with stable resonators. Negligible loss of adjustment through changes in the resonator construction or in the behavior of the laser medium produces significant changes in the laser output and spatial intensity distribution. Therefore, special constructions are generally necessary.
- The intensity distributions differ greatly in the near and far field.
- Unstable resonators show significant changes of the intensity distribution when they are used for processing reflecting working materials, since they react very sensitively to optical back reflections. The planned application has to be factored into the resonator construction in many cases.

Confocal resonators have proven to be functional since they deliver a collimated beam (Fig. 6.12). Both mirrors have a common focal point in confocal resonators

$$L = f_1 + f_2 = \frac{R_1}{2} + \frac{R_2}{2}, \tag{6.74}$$

and for the g parameters the relation

$$g_1 g_2 = \frac{1}{2}(g_1 + g_2) = \frac{(L - R_1)^2}{R_1^2 - 2LR_1}. \tag{6.75}$$

is valid. Instead of R_1, R_2 can also be selected as a free parameter. In Eq. 6.75, one can initially recognize that there is no stable confocal resonator, since $g_1 g_2$ is constantly larger than one when the denominator is positive. For the negative unstable branch, a focus lies in the resonator

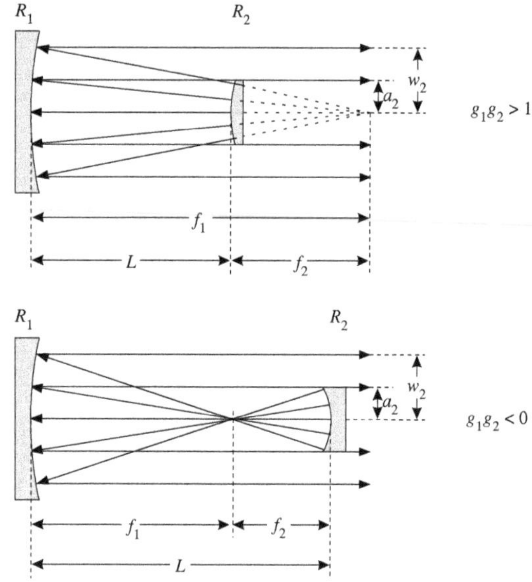

Fig. 6.12 Confocal resonators. *Above* is the positive branch, without focus in the resonator; *below* is the negative branch, in which the focus lies in the resonator

$$\begin{array}{lll} g_1g_2<0: & \frac{R_1}{2}\equiv f_1<L & \Rightarrow \text{ Focus in the resonator} \\ g_1g_2>1: & \frac{R_1}{2}\equiv f_1>L & \Rightarrow \text{ No focus in the resonator} \end{array} \tag{6.76}$$

For this reason, the positive branch is preferred, since a focus does not lie in the resonator and, therefore, the beam better covers the volume between the mirrors. This leads to a better exploitation of the active medium placed there. In addition, the beam intensity can be so high in the focus that the active medium there would be damaged.

A magnitude often used to describe the outcoupling for unstable resonators is the *magnification M*

$$M = \frac{w_2}{a_2} \tag{6.77}$$

w_2 Beam radius at the outcoupling mirror
a_2 Radius of the outcoupling mirror

It represents the enlargement of the beam radius as it passes through the resonator. In the case of the confocal resonators the following results:

$$M = \begin{cases} 2g_2 - 1 = \frac{1}{2g_1-1} = \frac{g_2}{g_1} & \text{for } g_1g_2 > 1 \\ 2g_1 - 1 = \frac{1}{2g_2-1} = \frac{g_1}{g_2} & \text{for } g_1g_2 < 0. \end{cases} \tag{6.78}$$

Using the magnification M, the beam losses of the unstable resonator can be very simply estimated. In the geometrical-optics approximation, the losses are given by the ratio of the beam cross section area and the mirror surface area. The loss factor δ follows for a resonator round trip

$$\delta = 1 - \tfrac{1}{M^2} \quad (M > 1)\,. \tag{6.79}$$

δ indicates the share of the beam output which is outcoupled out of the resonator. In geometrical-optics approximation, this share is emitted in a circular ring with the inner radius a_2 and the outer radius w_2.

6.5.1 Field Distribution of Unstable Resonators

The eigenmodes of unstable resonators fundamentally differ from the Hermite-Gaussian or Laguerre-Gaussian modes; therefore, they cannot even be represented by them approximately. Closed, analytic solutions of the wave equation for unstable resonators are not as yet known. Therefore, the field distribution in the resonator can only be gained by using numerical methods.

On the basis of several simplified assumptions, an approximate description of the emitted beam can, however, be derived. For the fundamental mode of unstable resonators with circular-symmetric outcoupling, a ring-formed near-field distribution originates in the geometrical optics approximation

$$E(r,0) = \begin{cases} 0 & r < a \\ E_0 & a < r < M \cdot a \\ 0 & M \cdot a < r \end{cases} \tag{6.80}$$

a Radius of the outcoupling mirror
M Magnification

The far-field distribution in the Fraunhofer approximation is given by the Fourier transform of this distribution. By these means, the intensity distribution in the far-field results in

$$I(\rho) = I(0)\left\{\frac{M^2}{M^2-1}\left(\frac{2J_1(\rho)}{\rho} - \frac{1}{M^2}\frac{2J_1(\rho/M)}{\rho/M}\right)\right\}^2 \quad \text{with} \quad \rho = 2\pi M\theta\frac{a}{\lambda}. \tag{6.81}$$

$J_1(\rho)$ Bessel function of the first type, first order
$\theta \approx \frac{r}{z}$ Beam angle in far field

Two limiting cases can be discussed here. When M is ≫1, then the outer radius of the ring in the near field is much larger than the inner radius. The intensity distribution in the far-field approaches the diffraction pattern of a simple aperture:

$$I(\rho) = I(0) \cdot 4\left(\frac{J_1(\rho)}{\rho}\right)^2 = I(0) \cdot \mathrm{AIRY}(\rho). \tag{6.82}$$

AIRY*(x)* Airy function

The first minimum of the distribution lies at

$$\theta = 0.51\frac{\lambda}{Ma}. \tag{6.83}$$

In the second limiting case, M is ≈ 1; the circular ring in the near field is then negligible. Then

$$I(\rho) = I(0) \cdot J_1(\rho)^2, \tag{6.84}$$

follows for the intensity distribution in the far field, and the minimum lies at

$$\theta = 0.38\frac{\lambda}{a}. \tag{6.85}$$

Figure (?) shows both limiting cases. In the second case, the full width at half maximum of the main maximum is significantly smaller, but the intensity in the secondary maxima is, however, higher. Figure 6.13 shows the numerically calculated near- and far-field distributions of an unstable resonator.

The differing field distribution in the near and the far-field demands precise setting of the working distance for many applications. A further difficulty is that, although the far-field distribution can be calculated from the near-field distribution, the near-field distribution must firstly be known, *including the phase information.* Since only intensities can be measured by direct means, the intensity distribution in focus cannot be reconstructed precisely enough from the near-field distribution. Therefore, the intensity distribution can only be monitored by measuring it directly in the focus. For high-performance lasers, this can, however, only be done with great difficulty.

6.6 Resonator Losses

The losses of a resonator determine the sharpness of the resonance and, thus, the quality of the emitted laser radiation. In addition, the threshold for laser activity (cf. Sect. 6.2.2) and the laser output power are dependent upon the resonator losses. The resonator losses, therefore, represent an important parameter of a laser, one which has to be incorporated when a laser system is being designed and constructed.

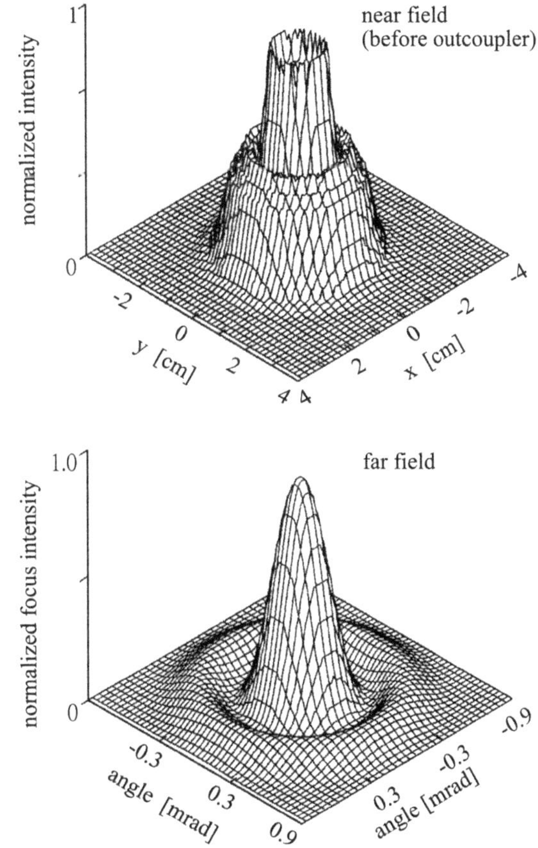

Fig. 6.13 Numerically calculated near- and far-field distribution of an unstable, confocal resonator ($R_1 = -14$ m, $R_2 = 28$ m, $M = 2$). The inner part of the near-field distribution is reflected at the resonator mirror; only the outer ring is outcoupled

The resonator losses contain all the losses that the radiation field experiences during a round trip in the resonator. At first, the *outcoupling losses* and the *dissipative losses* have to be distinguished. The outcoupling losses arise from the outcoupling of the laser beam out of the resonator, which means they describe the part of the output, conveyed to the application and provided by the laser system. They represent a loss only in relation to the resonator; in relation to the laser application, in contrast, they represent the actual system output. The dissipative losses, however, are lost in the form of heat or scattered light, without being able to be used. The dissipative losses, therefore, have to be minimized, whereas the outcoupling losses have to optimized in view of the necessary system properties

- Strong outcoupling: high-output power, low-resonator quality, which means large bandwidth, or
- Low outcoupling: low-output power, high-resonator quality and narrow bandwidth.

The optimization of the outcoupling losses is discussed in Sect. 8.4.3. The dissipative losses are essentially made up of

- Diffraction losses, determined by the finite mirror dimensions, and
- Technically determined losses
  - Absorption losses from the reflection at the mirrors,
  - Scattering losses through unevenness of the mirrors, and
  - Losses due to maladjustment of the resonator which results in light leaving the resonator.

These loss mechanisms will be dealt with in the next section.

6.6.1 Diffraction Losses

In general, the radiation circulating in the resonator is not completely caught by the resonator mirror and reflected again. A share of it passes through the mirror outside the reflecting surface and thus leaves the resonator. In the case of stable resonators, the output lost this way completely represents a dissipative loss since this radiation is not used. With unstable resonators, however, this process is used as an outcoupling mechanism, and the largest share of the radiation passing through the mirror laterally forms the outcoupled laser beam.

The cutoff of the intensity through the mirror leads to the beam distribution only being approximately described by Hermite-Gaussian modes. The approximation is the more precise, the smaller the beam radius w is on the mirror in relation to the mirror radius a. The condition

$$w \ll a \tag{6.86}$$

is the easiest to fulfill for symmetric confocal resonators. In concentric and plane-plane resonators, the beam radii grow constantly, however, until they are limited by the diffraction losses. For the field distributions of these resonators, the Hermite-Gaussian modes do not represent an appropriate approximation.

Calculating the diffraction losses demands that the diffraction problem of the optical resonator to be solved. In general, this is only possible using numerical methods. For several simple cases, however, analytic approximations are known. For example, the following is valid for the diffraction losses of the fundamental mode of symmetric-confocal resonators according to Slepian

$$\begin{aligned} \delta_B &\approx 1 - 35.5\sqrt{N_F}\mathrm{e}^{-4\pi F} \quad (N_F > 0.5) \quad \text{quadratic mirror} \\ \delta_B &\approx 1 - 16\pi^2 N_F \mathrm{e}^{-4\pi F} \quad (N_F > 0.5) \quad \text{circular mirror} \end{aligned} \tag{6.87}$$

δ_B Relative diffraction losses per circulation
N_F Fresnel number

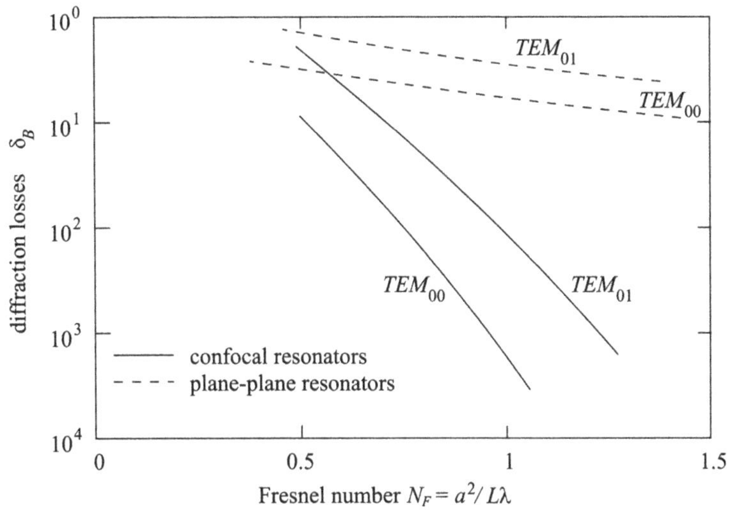

Fig. 6.14 Diffraction losses of the two lowest transversal modes for confocal, or plane-plane resonators with circular aperture as a function of the Fresnel number a is the mirror radius, L the resonator length and λ the wavelength

In Fig. 6.14 the diffraction losses according to Eq. 6.87 are plotted against the Fresnel number. A detailed treatment of the diffraction losses of different resonators and modes can be found in Fox and Li as well as in Boyd and Gordon.

6.6.2 Absorption and Scattering at the Mirrors

The laser radiation striking the mirror is not completely reflected, nor transmitted. A share is absorbed in the mirror material and leads to a heating of the mirror. For this absorption loss, the following is generally valid:

$$\delta_A = 1 - R - T. \tag{6.88}$$

R Reflexion coefficient
T Transmission coefficient

The numerical value of δ_A depends upon the mirror material and the wavelength of the laser radiation. Appropriate coatings can increase the reflectivity of the mirror significantly.

In particular, for long folded resonators, the absorption losses can lead to a significant reduction of efficiency. In addition, strong heating of the mirror can lead to a change in its geometry on account of thermal expansion and, in the worst case, to its destruction. In high-performance operation, therefore, the mirrors have to be cooled (Fig. 6.15).

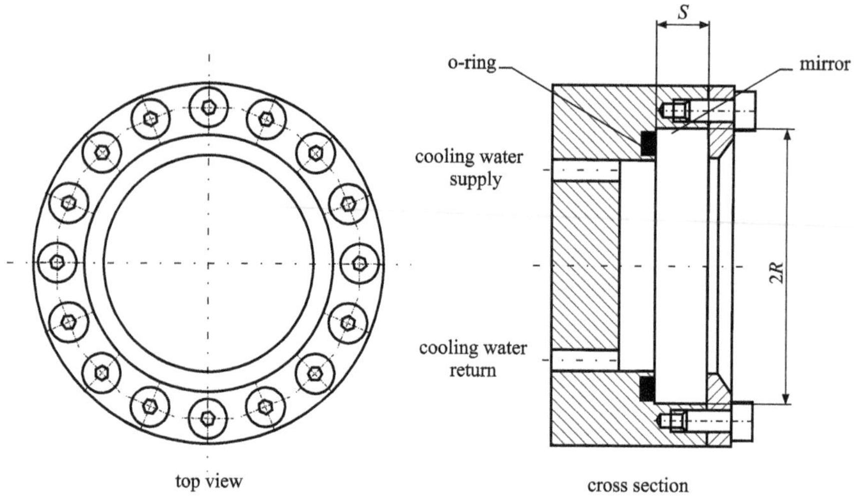

Fig. 6.15 Construction of a water-cooled copper mirror for use in high-performance laser plants. Important is that the mirror rests evenly on the front mounting ring, so that the cooling water pressure does not lead to poor adjustment

For CO_2 lasers with a wavelength of 10.6 μm, *copper mirrors* are used in practice. At this wavelength they absorb approximately 1 % of the beam power hitting them; they represent a compromise between good reflection and quick thermal dissipation while also being economical to produce.

Due to unavoidable, microscopic roughness of the mirror surfaces, a share of the reflected radiation is scattered. Impurities, for example dust, can strengthen this effect. As a standard quality for optic surfaces, roughness values of less than $\lambda/10$ are used in laser engineering. Thus, surface-based scattering losses can be reduced to less than 1 %.

Dielectric mirrors

Using *dielectric mirrors*, absorption losses of less than 0.2 % can be reached in the visible as well as in near-infrared and near-ultraviolet spectral range. Dielectric mirrors are produced by applying one or many $\lambda/4$ layers on a glass substrate. The refractive index, n_{layer}, has to be larger than the refractive index of the glass, n_{glass}. The part of the radiation reflected on topside of the layer undergoes a phase jump of π because of the transition to an optically denser medium ($n_{\text{air}} < n_{\text{layer}}$). On the other hand, the part of the wave reflected at the interface between layer and glass does not undergo a phase jump due to $n_{\text{layer}} > n_{\text{glass}}$, covers, however, a path longer by $\lambda/2$. In total, this leads to a constructive interference of both partial waves and thus to an enhanced reflection.[7]

[7]According to the same principle, an antireflection coating for an optical element can also be designed. For this, only $n_{\text{layer}} < n_{\text{glass}}$ has to be selected. The additional phase jump at the reflection at the inner side of the layer then leads to a destructive interference of the reflected partial waves.

What needs to be taken into account is that the layer applied can be optimized in its thickness only for one wavelength, which means the reflectivity of dielectric mirrors is strongly dependent upon wavelength. An example of a dielectric mirror dimensioned for Nd:YAG laser radiation cannot, in general, be used for other laser types. The disadvantage of dielectric mirrors is primarily the higher manufacturing costs.

6.6.3 Misalignment

A *misalignment* of a resonator mirror, which means a resonator geometry that minimally deviates from the ideal, also leads to an increase of resonator losses. An important misalignment is the *tilt* of a mirror with respect to the axis perpendicular to the resonator axis (Figure 6.16). This leads to the beam axis being tilted more and more with respect to the resonator axis with multiple reflections: the beam moves outwards. Through this, a larger share of the beam leaves the resonator at the side of the mirrors and is lost.

The plane-plane resonator with two plane mirrors exhibits the largest sensitivity with respect to the tilt of a mirror. The losses caused by misalignment δ_J are dependent upon the tilt angle ε and can be approximated by the expression

$$\delta_J = \frac{16}{3}\pi^2 N_F \frac{L}{\lambda}\varepsilon^2. \tag{6.89}$$

δ_J Misalignment loss
N_F Fresnel number of the resonator
L Resonator length
λ Wavelength

In order to reduce the misalignment losses to the same order of magnitude as the diffraction losses, according to this formula, the tilt angle ε has to be smaller than the diffraction angle:

$$\varepsilon < \theta_B = \tfrac{\lambda}{2a} \quad \Rightarrow \quad \delta_J < \delta_B\,. \tag{6.90}$$

This condition can be formulated in such a way that the difference of the distances between the mirror edges has to be smaller than a wavelength:

$$s = L_1 - L_2 < \lambda. \tag{6.91}$$

Such adjustment precision can only be attained with great difficulty in practice, especially since it has to be maintained during operation. In addition to the high diffraction losses, this is a further reason for planar resonators not finding wider use in laser engineering.

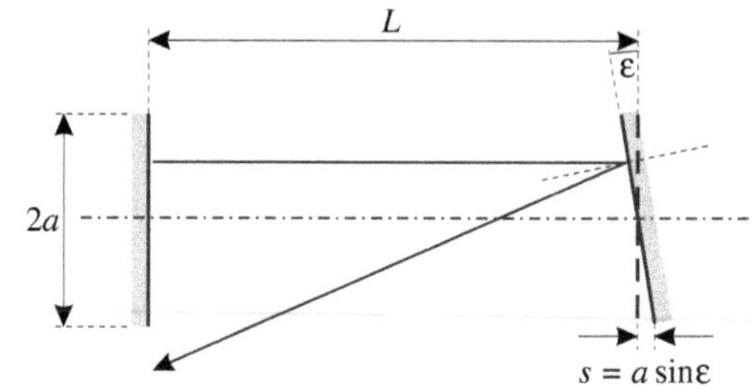

Fig. 6.16 Schematic representation of losses due to the misalignment of a mirror

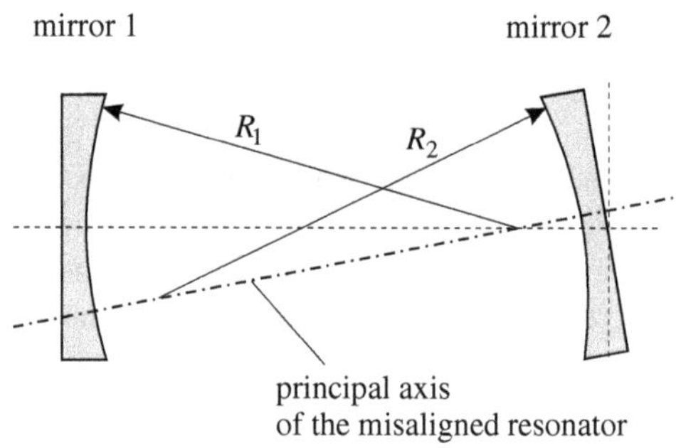

Fig. 6.17 With a spherical resonator, the tilting of the mirror causes the optical axis to reorient: the optical axis always runs through the center point of the mirror curvatures

Spherical resonators are far less sensitive to mirror tilt: with the tilt angle ε, only the position of the resonator axis changes, determined by the center of curvature of the mirror. The beam does not leave the resonator, as with planar resonators, since the beam axis remains stable even with multiple reflections (Fig. 6.17). In this case, additional losses only arise by the beam axis being shifted closer to the edge of the resonator mirror and thus cutting off a larger share of the beam.

By contrast, spherical resonators react with a great sensitivity to changes of the g parameter,

$$g_{1,2} = 1 - \frac{L}{R_{1,2}}, \tag{6.92}$$

$R_{1,2}$ Curvature radii of the resonator mirrors,

which means to the changes in the resonator length L or to the radii of curvature of the mirrors $R_{1,2}$. The waist radius and the divergence angle of the resonator modes can vary strongly depending on the g parameters, especially with resonators that lie on the border of stability region (cf. Fig. 6.8). Resonators with lowest sensitivity to misalignment lie, therefore, at

$$g_1 = g_2 = \pm\frac{1}{2} \tag{6.93}$$

in the middle of the stability region.

To compare the misalignment sensitivity of different resonators to each other, Eq. 6.89 is put into a general form, as follows:

$$\delta_J = C_R N_F \frac{L}{\lambda} \varepsilon^2. \tag{6.94}$$

C_R Constants dependent upon resonator type

The constant C_R expresses the misalignment sensitivity of the respective resonator type. It varies in the area

$$\begin{aligned} & C_R \approx 0.7: && \text{resonator with } g_{1,2} = 1/2 \\ \text{until} \quad & C_R \approx 50: && \text{planar resonator} \end{aligned},$$

which means the planar resonator exhibits approximately 70 times higher misalignment sensitivity at the same Fresnel number.

The static and dynamic stability of the resonator construction are the prerequisites for a high-adjustment precision and thus for low resonator losses. Larger losses always mean a reduced quality of the laser radiation. *Mechanical oscillations* of the resonator during operation can, for example, lower the mode quality and the mean output. Such oscillations are generated by other laser components, above all by the systems used for pumping, or they are transferred by other devices in the form of acoustic waves via the building floor. Therefore, special emphasis is placed on damping oscillations when the resonator is engineered.

Other causes for misalignment losses of the resonator are the *thermal expansion* or the *elastic deformation* through mechanical loading of the construction elements. Hence, substances with low heat expansion coefficients are preferred as materials for construction. Stable pipe and carrier constructions are used to counteract elastic deformation (Fig. 6.18).

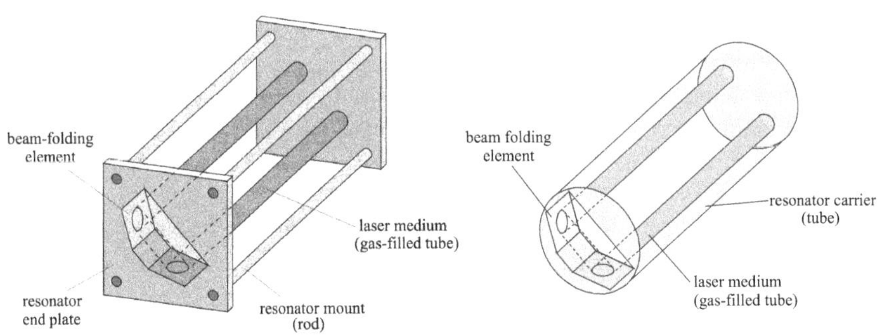

Fig. 6.18 *Left* Rod construction of the resonator carrier. Discharge path and head with deflecting mirrors of this folded resonator are held in place by massive end plates, which are connected by outer rods. *Right* Pipe construction with an enclosing pipe as a resonator carrier

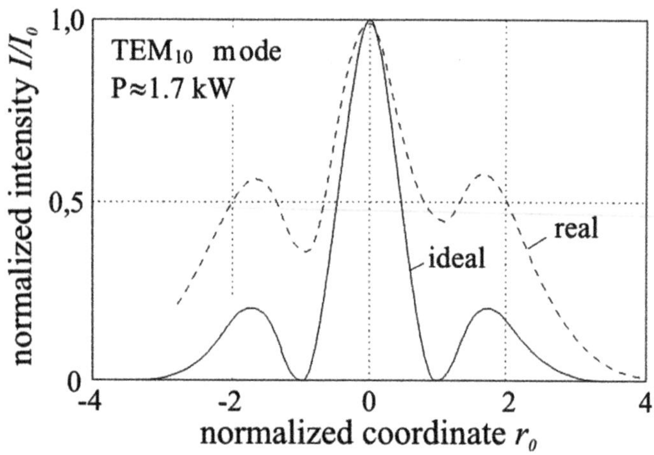

Fig. 6.19 Influence of the laser medium on the transversal intensity distribution in the resonator: higher amplification in the outer zones due to the saturation in the axial area

6.6.4 Influence of the Laser Medium

When a resonator is dimensioned, it is not sufficient to only look at the characteristics of the empty resonator. the propagation of the radiation inside the the resonator is, in some cases crucially, influenced by the laser medium. The effects are very specific for the respective medium; therefore, Chap. 10 gives a more detailed representation in connection with the corresponding laser system.

In general, the laser medium can lead to a change of the *optical geometry* of the resonator. The refractive index of the medium can depend strongly on the temperature (e.g., for Nd:YAG lasers) or on the excitation density caused by the pump mechanism (e.g., semiconductor lasers). Corresponding to this dependency, a refractive index profile forms perpendicular to the resonator axis, which, for example, acts like a lens (thermal lens in Nd:YAG laser, cf. Sect. 10.9.3). This additional lens has to be taken into account when the radii of curvature of the mirrors and resonator length are calculated.

In solid-state lasers, the high-radiation intensity in the resonator can also lead to the laser medium becoming *birefringent*; this means the refractive index depends on the polarization direction of the radiation. The consequence of this is that a share of the radiation leaves the resonator due to refraction.

With gas lasers high speeds of the circulating gas lead to *turbulence* and *density fluctuations*, which, in turn, lead to diffraction of the light and thus to additional losses.

Further fundamental influences of the gain medium on the beams field distribution have their causes in the laser-typical high *nonlinearity* of the amplification. Due to the saturation of the amplification at high intensities, the raising and trailing edges of the intensity distribution are raised relative to the maximum of the intensity distribution, and the zero positions of the ideal distribution are filled (Fig. 6.19).

Temporal and spatial inhomogeneity in the gain profile lead to fluctuations in the intensity distribution and to a mixing of the resonator modes.

Such influences can only be calculated using extensive numerical simulations of the resonator and laser medium.

Chapter 7
Interaction of Light and Matter

In the previous chapters the propagation of electromagnetic waves has been described in different ways. Chapter 5 already focused on the propagation of well confined light beams: laser beams. With the optical resonator Chap. 6 finally addressed the first main element of a laser.

However, the heart piece of the laser, the *active medium*, generating and amplifying the laser radiation, has not been addressed yet, and the influence of matter on the propagation of radiation has been considered only in the form of macroscopic optical parameters such as the index of refraction n and the absorption index α. The microscopic structure of matter and the mechanisms for its interaction with radiation were not required to model the propagation of radiation.

In this chapter this gap will be closed. The inner structure of matter and the fundamental mechanisms of its interaction with radiation will be described with the scope to explain the generation of laser radiation in the active medium. For details beyond this scope we refer to the respective specialized literature.

Whether gaseous, liquid or solid, matter consists of *atoms*. There are many different types of atoms, distinguished by their mass and the electrical charge of their nucleus. The different types of atoms, or *elements*, are arranged in the *periodic system of the elements* according to their *nuclear charge* and the structural resemblance of their *electron shell*.

Over time, many different models have been developed to describe the inner structure and the resulting properties of atoms. Table 7.1 gives an overview of the atomic models in modern times. The atomic models have been evolved continuously to comprehend newly discovered phenomena. In particular, the development of *quantum mechanics* by SCHROEDINGER and HEISENBERG at the beginning of the twentieth century lead to a revolution in the understanding of the inner structure of atoms and their interaction with radiation.

However, it is not always necessary to employ the most advanced and complex model. Many properties of matter can already explained with comparably simple, classical atomic models.

R. Poprawe et al., *Tailored Light 1*, RWTHedition,
DOI 10.1007/978-3-642-01234-1_7

Table 7.1 Overview of the atomic models of modern times, in the order of their development

Symbol	Model	Achievement as a model for interaction with radiation
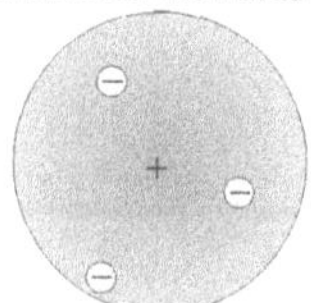	Thomson Model: Electrons with negative charge embedded in positively charged body	Atom as a dipole (Lorentz Model): Dispersion, classical attenuation of radiation
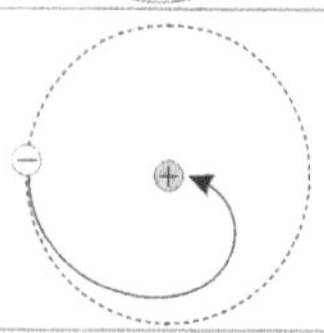	Rutherford Model: Mass of the atom concentrated in the (postively charged) atomic nucleus	Inherent instability: electrons continuously emit radiation and fall into the atomic nucleus
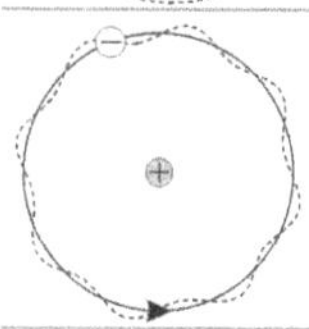	Bohr Sommerfeld Model: Bohr's Postulates; electrons on closed orbits subject to quantum conditions	Elimination of the radiative instability and prediction of the atomic line spectrum as a consequence of discrete energy levels of the atom
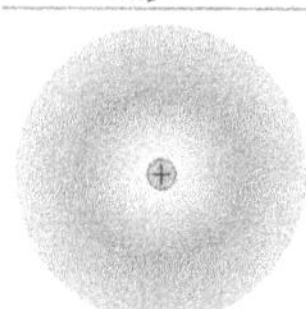	Quantum mechanical atom: Quantum mechanical description of the electrons in the Coulomb potential of the nucleus	Energy (wavelength) and relative intensity of the spectral lines; absorption and stimulated emission
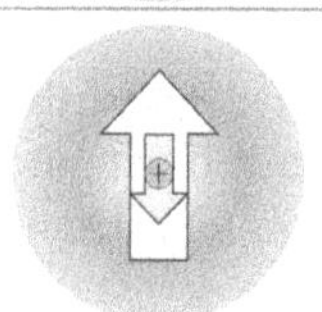	Quantum mechanical corrections: relativistic effects; electron spin; nuclear spin	Adjustments to the spectral lines; fine structure and hyperfine structure of the energetic levels of the atom
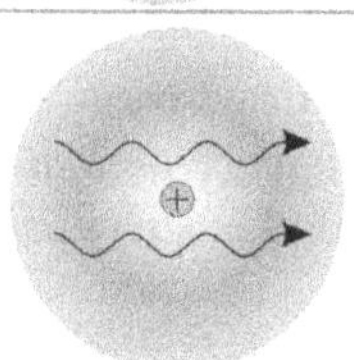	Quantum Electrodynamics: Interaction with the quantized radiation field; vacuum fluctuations of the electromagnetic field	Spontaneous emission and natural linewidth; Lamb shift of energy levels

7.1 Absorption and Emission of Light—Spectral Lines

Already in the early nineteenth century a number of dark lines were discovered in the spectrum of the sun. These lines are called *Fraunhofer lines* after their discoverer Joseph von Fraunhofer. These lines are produced by the absorption of the light at specific wavelengths in the chromosphere of the sun and the atmosphere of the earth. Kirchhoff and Bunsen investigated 1860 the absorption spectra of

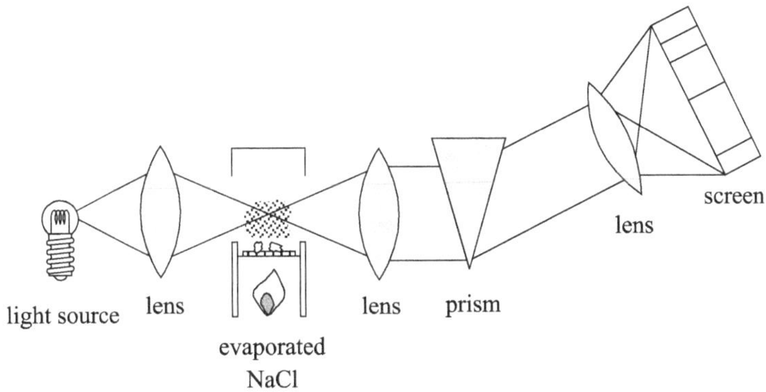

Fig. 7.1 Schematic representation of an experiment for the determination of the absorption spectrum of sodium. The screen shows a continuous spectrum with *dark lines*. These are the absorption lines of sodium

sodium and other elements. In the following period of time the spectroscopy, as the determination of spectral lines is called, developed to one of the most important tools for the development of new theories for the description of atoms and matter.

Spectroscopic investigations can be performed in two fundamentally different configurations. For the *absorption spectroscopy* that was exclusively used in the early years, the light of a broadband light source transmits the matter to be investigated. With a prism the spectral components of the light are spatially resolved and projected on a screen (Fig. 7.1). The *absorption lines* are represented by dark lines in the continuous spectrum of the light source.

For the emission spectroscopy, a light source containing the evaporated, gaseous material to be investigated is used, e.g. a sodium vapor lamp (Fig. 7.2). In this case the spectrum only comprises of the emission lines of this particular material.

The spectroscopic experiments yield the following statements that are important for the understanding of the atomic structure (Fig. 7.3):

- Each type of atom has characteristic emission and absorption lines. Each atom can only absorb and emit light at specific and discrete wavelengths.
- Wavelength and intensity of these lines are identical in the emission and absorption spectra. However, not all lines occur in both spectra; typically the emission spectrum contains significantly more lines than the absorption spectrum.
- For each type of atom exists a cutoff wavelength, below which no emission or absorption lines exist anymore. The spectra are not limited toward long wavelengths.

The attempt to explain these experimental results drove the development of new atomic models and finally led to a completely new model of the microcosm: The *quantum mechanics*.

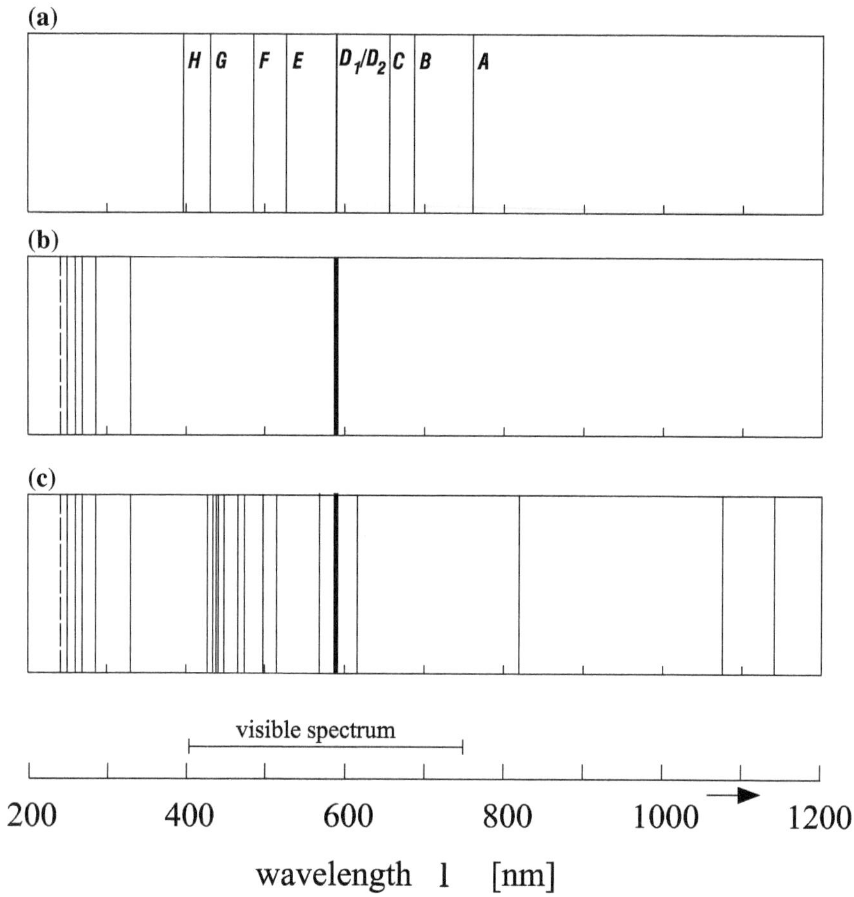

Fig. 7.2 **a** The Fraunhofer lines in the absorption spectrum of the sun. The lines A and B are produced by the absorption in the atmosphere of the earth, the others by absorption in the chromosphere of the sun. **b** Absorption spectrum of sodium. The most intense lines are the D_1 and D_2 lines that can be observed in the absorption spectrum of the sun, too. **c** Emission spectrum of sodium

7.2 The Dipole Model

The classical atomic models proved to be insufficient for the interpretation of emission and absorption spectra of the atoms. Obviously, the internal structure of the atoms cannot be described correctly with the means of classical physics. Nevertheless, the classical atomic models lead to at least one conclusion of fundamental importance: Atoms contain positive and negative electric charges, and the arrangement of these charges can be influenced by external forces.

Based on this knowledge a model can be developed that reproduces some of the fundamental electromagnetic properties of the matter. Assuming each atom consists

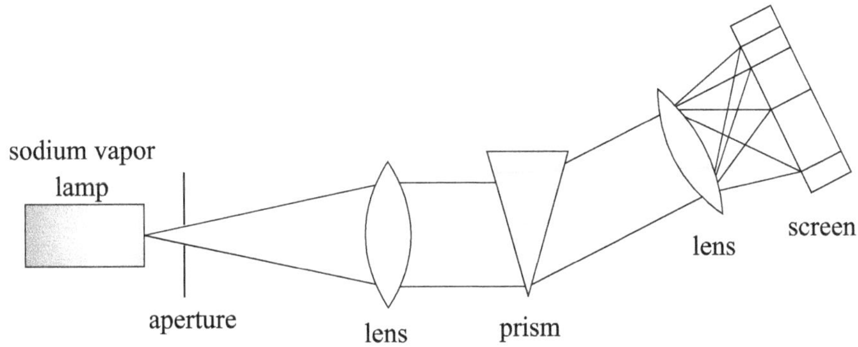

Fig. 7.3 Emission spectroscopy of sodium. The emission lines of sodium are visible on the screen as bright lines

of positively and negatively charged elements, an external electrical field will lead to a spatial separation of these charges, with the degree of separation being proportional to the electric field strength. Thus, a *dipole* is generated (Fig. 7.4). Is the atom subjected to an oscillating electric field, then an oscillating dipole is induced that emits radiation itself.

7.2.1 The Lorentz Model

The *dipole moment* of a dipole consisting of a positive charge $+q$ and a negative charge $-q$ is given by

$$\vec{p} = q \cdot \vec{d}, \tag{7.1}$$

where $\vec{d}$ is the displacement vector pointing from the negative charge to the positive charge[1] (Fig. 7.4). In the absence of external fields the charges maintain the equilibrium distance d_0 to each other. The force of an electric field acting on a charge q is

$$\vec{F} = q \cdot \vec{E}.$$

In an external, oscillating field

$$\vec{E}(t) = \vec{E}_0 \mathrm{e}^{-i\omega t}$$

[1] In general the dipole moment is defined as $\vec{p} = \int \mathrm{d}\vec{x}\, \rho(\vec{x}) \cdot \vec{x}$, with the electric charge density ρ. For approximately point-shaped charges the integral can be reduced to the formula according to Gl. 7.1.

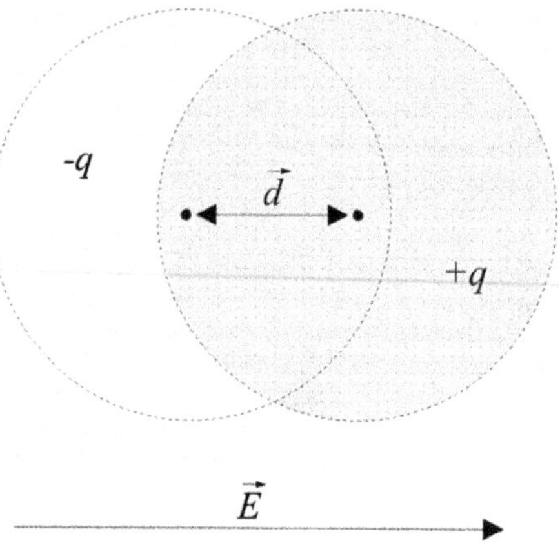

Fig. 7.4 In the electrical field positive and negative charges are separated: a dipole is generated

therefore the charges of the dipole oscillate. This oscillation can be described by the differential equation

$$\ddot{\vec{d}} + b\dot{\vec{d}} + \omega_0^2\left(\vec{d} - \vec{d}_0\right) = \frac{q}{m}\vec{E}_0 e^{-i\omega t}. \tag{7.2}$$

b damping coefficient
ω_0 eigenfrequency of the dipole
m mass of the oscillating charge.

The eigenfrequency and the damping of the oscillation depend on the elastic force between the charges of the dipole. The solution of this differential equation is

$$\vec{d}(t) = \vec{d}_0 + \vec{d}_1 e^{-i\omega t}, \quad \vec{d}_1 = \frac{q\vec{E}_0}{m}\frac{1}{\omega_0^2 - \omega^2 - i\omega b}. \tag{7.3}$$

From this results the dipole moment

$$\vec{p}(t) = q \cdot \vec{d}(t) = q\vec{d}_0 + q\vec{d}_1 e^{-i\omega t} \equiv \vec{p}_0 + \vec{p}_1(t), \tag{7.4}$$

where the oscillating part is proportional to the electrical field:

$$\vec{p}_1(t) = \alpha \cdot \vec{E}(t), \quad \alpha(\omega) = \frac{q^2}{m\left(\omega_0^2 - \omega^2 - i\omega b\right)} \tag{7.5}$$

α is the *polarizability* of the atom. The macroscopic *polarization* of a medium consisting of N atoms is simply the sum over the dipole moments of all the atoms:

$$\vec{P} = N \cdot \vec{p} = Nq \cdot \vec{d}. \tag{7.6}$$

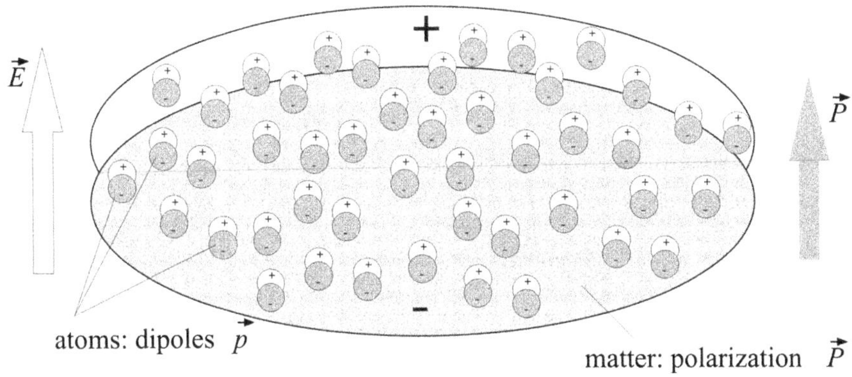

Fig. 7.5 By an external electrical field atomic dipoles are induced. This leads to a polarization of the whole medium

$\vec{P}$ *Polarization of the medium*

In this way the microscopic displacement of the dipole charges d gets correlated to a macroscopic material property. This model is called the *Lorentz model* of the interaction between light and matter (Fig. 7.5).

7.2.2 The Complex Index of Refraction

The linearized relation between the macroscopic polarization and the electric field was already given in Sect. 3.2.4:

$$\vec{P} = \varepsilon_0 \chi(\omega) \vec{E} \equiv \varepsilon_0 (\varepsilon(\omega) - 1) \vec{E}. \tag{7.7}$$

ε_0 vacuum permittivity
$\chi(\omega)$ electrical susceptibility
$\varepsilon(\omega)$ dielectric function

This relation is expressed either using the electrical susceptibility or the dielectric function. Generally, both functions are tensors: The electric field vector $\vec{E}$ and the polarization vector $\vec{P}$ do not have to be parallel. By comparing Eq. 7.7 with the Eqs. 7.6 and 7.5 the electric susceptibility in the dipole approximation can be derived:

$$\chi(\omega) = \frac{1}{\varepsilon_0} N \alpha(\omega) = \frac{N q^2}{m \varepsilon_0} \frac{1}{\omega_0^2 - \omega^2 - i\omega b}. \tag{7.8}$$

In this case the electrical susceptibility is a complex and scalar function of the angular frequency ω. Also from Sect. 3.2.4 the relation between dielectric function and refractive index is known:

$$\tilde{n}^2 = \varepsilon(\omega) = \chi(\omega) + 1. \tag{7.9}$$

$\tilde{n}$ complex refractive index

Since the susceptibility is complex, also the refractive index has to be complex. The interpretation of the *complex refractive index* gets obvious when looking at a wave propagating in the described medium:

$$E(z) \sim \exp(i\tilde{n}kz) = \exp(i\tilde{n}_r kz)\exp(-\tilde{n}_i kz), \quad \tilde{n}_r = \Re(\tilde{n}), \tilde{n}_i = \Im(\tilde{n}). \tag{7.10}$$

While the real part of the complex refractive index corresponds to the real refractive index, its imaginary part describes the attenuation or *absorption* of the wave while penetrating into the medium:

$$\Re(\tilde{n}) = n, \quad \Im(\tilde{n}) = \kappa. \tag{7.11}$$

n real refractive index
κ absorption index

Splitting the complex index of refraction according to Eq. 7.9 into real and imaginary part leads to the following relations between real refractive index, absorption index and susceptibility:

$$\tilde{n}^2 = n^2 - \kappa^2 + 2in\kappa \Rightarrow \quad \begin{aligned} n^2 - \kappa^2 &= 1 + \Re(\chi), \\ 2n\kappa &= \Im(\chi) \end{aligned} \tag{7.12}$$

In some cases, e.g. for thin gases, the approximation $\tilde{n} \approx 1$ is valid. In this case real and imaginary part of the susceptibility can be related directly to the refractive index and absorption index:

$$\begin{aligned} n(\omega) &\approx 1 + \frac{1}{2}\Re(\chi) = 1 + \frac{q^2 N}{2\varepsilon_0 m} \frac{\omega_0^2 - \omega^2}{\left(\omega_0^2 - \omega^2\right)^2 + \omega^2 b^2}, \\ \kappa(\omega) &\approx \frac{1}{2}\Im(\chi) = \frac{q^2 N}{2\varepsilon_0 m} \frac{\omega b}{\left(\omega_0^2 - \omega^2\right)^2 + \omega^2 b^2}. \end{aligned} \tag{7.13}$$

In Fig. 7.6 these relations are shown as a function of the frequency of the external electrical field ω.

7.2.3 The Dispersion Relation

The dependency of the refractive index on the frequency is generally referred to as *dispersion*. When discussing the function $n(\omega)$ three areas are distinguished:

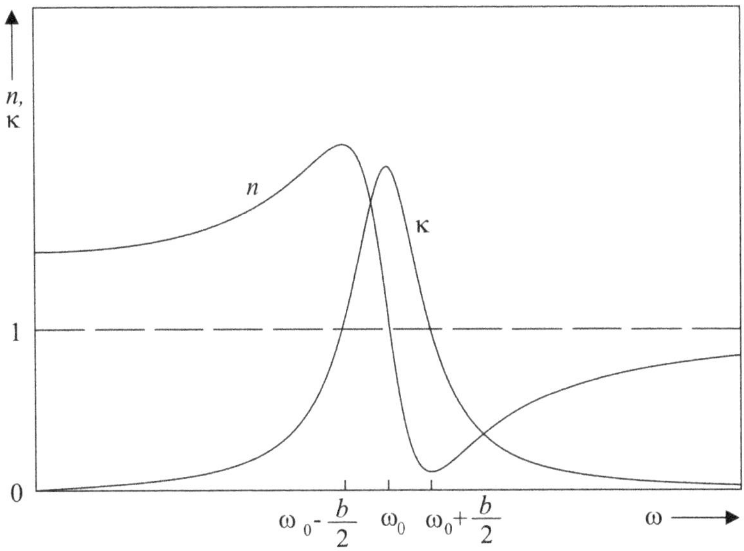

Fig. 7.6 Refractive index n and absorption index κ as a function of the frequency ω, according to the Lorentz model

- For frequencies far below the resonance frequency, the limit $\omega \to 0$ can be taken:

$$\lim_{\omega \to 0} n(\omega) = 1 + \frac{q^2 N}{2\varepsilon_0 m \omega_0^2}.$$

  Thus, the refractive index approaches a value that depends on material parameters only, but is larger than 1.
- For ω much larger than ω_0 follows

$$\lim_{\omega \to \infty} n(\omega) = 1\,,$$

  thus the refractive index approaches 1 for frequencies significantly larger than the resonance frequency.
- Close to the resonance, $\omega \approx \omega_0$, the refractive index can be approximated as

$$n(\omega) \approx 1 + \frac{q^2 N}{2\varepsilon_0 m} \frac{\omega_0^2 - \omega^2}{\omega_0^2 b^2}$$

by substituting ω by ω_0 in the denominator. At $\omega = \omega_0$ the refractive index is 1, and for $\omega > \omega_0$ the refractive index is always smaller than 1.

Generally, atomic systems have several resonance frequencies. The dispersion curve then consists of several sections with the discussed behavior (see Fig. 7.7).

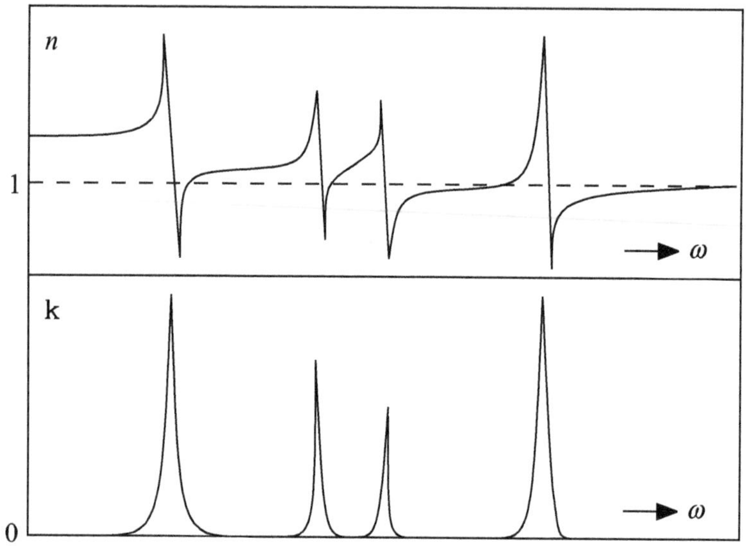

Fig. 7.7 Schematic representation of the absorption index κ and the refractive index n for an atomic system with several resonance frequencies

For historic reasons the areas where the refractive index grows with increasing frequency are called areas with *normal dispersion*, whereas the other case *is called anomalous dispersion*. Normal dispersion is observed for the refraction of visible light on most glasses: Red light (longer wavelength and lower frequency) is refracted less than blue light (shorter wavelength and higher frequency).

7.2.4 Absorption

Now we look at the absorption index $\kappa(\omega)$ as given in Eq. 7.13. The absorption index approaches zero for frequencies well below and well above the resonance frequency:

$$\lim_{\omega \to 0} \kappa(\omega) = 0, \quad \lim_{\omega \to \infty} \kappa(\omega) = 0 .$$

In the vicinity of the resonance frequency, $\omega \approx \omega_0$, the absorption index takes on its maximum. The exact position of the maximum is shifted against the resonance frequency by the damping coefficient b; it is approximately where the denominator in Eq. 7.13 takes on its minimum:

$$\frac{\mathrm{d}}{\mathrm{d}\omega}\left[\left(\omega_0^2-\omega^2\right)^2+b^2\omega^2\right]=-4\omega\left(\omega_0^2-\omega^2\right)+2b^2\omega=0 \quad\Rightarrow\quad \omega_{\max}^2=\omega_0^2+\frac{b^2}{2}. \tag{7.14}$$

In case of a small damping coefficient b, the maximum absorption index can be approximated by

$$\omega_{\max}\approx\omega_0: \quad \kappa_{\max}\approx\kappa(\omega_0)=\frac{q^2N}{2\varepsilon_0 m}\frac{1}{\omega_0 b}. \tag{7.15}$$

Thus, the maximum absorption index is inversely proportional to the damping coefficient: A weak damping leads to a resonance line with strong absorption.

As shown in Eq. 7.10, the absorption index describes the reduction of the electric field amplitude for a wave traveling in the medium:

$$E(z)\sim\exp(-\kappa kz),$$

The intensity of the wave is proportional to the square of the field amplitude; thus

$$I(z)=I_0\exp(-2\kappa kz)=I_0\exp(-\alpha z), \quad \alpha=2k\kappa. \tag{7.16}$$

α absorption coefficient

This is *Lambert Beer's Law* for the reduction of the intensity in an absorbing medium. The validity of this law is easily proved experimentally (Fig. 7.8).

For an *amplifying laser medium* just the reverse process is needed: The intensity has to increase while traveling in the medium. Obviously, the absorption coefficient α has to be negative to achieve this effect. In this case an *amplification coefficient* or *gain coefficient* is defined:

$$g=-\alpha>0 \tag{7.17}$$

g gain coefficient

In the current model the absorption index κ would have to be negative to achieve a positive amplification coefficient. However, according to Eq. 7.13 this is not possible since the damping coefficient b is, for fundamental reasons, always positive.[2] This shows that, although this simple, classical physics model is sufficient to explain

[2]In the classical picture, the attenuation describes the reduction of the oscillation amplitude as a consequence of a dissipative energy loss, e.g. due to mechanical friction. These processes cannot be reversed to result in an energy gain (2nd law of thermodynamics). Thus the damping coefficient is always positive.

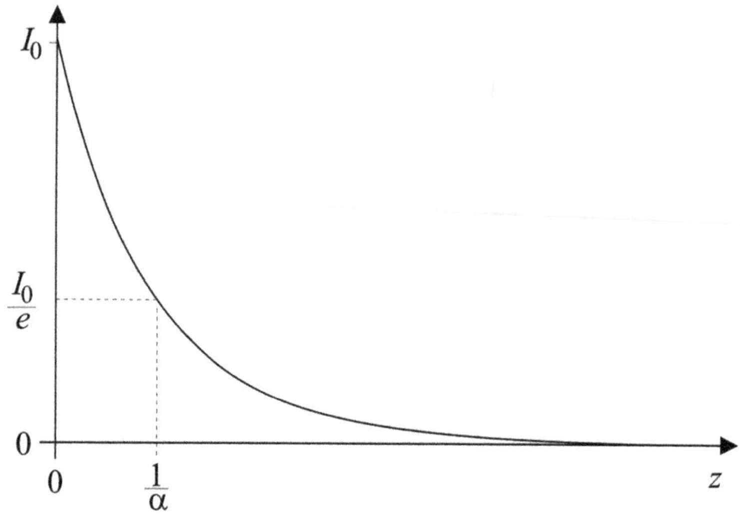

Fig. 7.8 Lambert Beer's Law: exponential decay of the intensity in a medium with the absorption α

fundamental refraction and absorption phenomena, it is inadequate to explain amplification in a laser medium. For this a *quantum physical* model is required.

7.3 Quantum Physics, Photons and Rate Equations

All models for the interaction of radiation with matter that were based on classical mechanics and electrodynamics proved to be insufficient in many respects. Many experimental findings, such as the line spectra, the X-ray emission and the photo effect could not be explained. MAX PLANCK was the first to postulate the quantization of energy when he formulated his law on the spectral distribution of radiation in a thermal equilibrium (*Planck's radiation law*) in 1900. ALBERT EINSTEIN presented in 1905 an explanation of the *photo effect* based on the absorption of light quantae or *photons*. For this explanation he received the Nobel prize in 1921. This was the starting point for the development of quantum physics that revolutionized the world view of physics since then.

The two most fundamental statements of quantum physics are:

1. There is no fundamental difference between particles and waves. Depending on the measurement, particles can exhibit wave characteristics and, vice versa, waves can behave like particles.
2. Certain phenomena and measurable values are fundamentally quantized. They occur only in certain steps (quantae) and not as continuous variables as in classical physics.

These two fundamental characteristics have widespread consequences for the interpretation of phenomena in microscopic physics.

Quantum mechanics deals with the quantum physical description of particles, taking into account their respective wave characteristics, while waves and fields are described with the methods of classical physics. The theory of quantum mechanics has been independently developed by WERNER HEISENBERG and ERWIN SCHROEDINGER, in 1925 and 1926 respectively. SCHROEDINGER derived the equation that fundamentally describes the evolution of quantum mechanical *wave functions* in time and space: The *Schroedinger equation*.

The *quantum field theory* then extends the quantum physical model to fields and leads to a symmetric description of particles and fields. In fact, field and particles are not distinguished anymore. Only the type of measurement determines whether particle or wave characteristics become evident. The fundamentals of the quantum field theory have been developed by FEYNMAN, SCHWINGER and TOMONAGA around 1950 for the specific case of *quantum electrodynamics*, i.e., the quantum physical description of electromagnetic fields and waves.

For a well-founded representation and explanation of quantum physics as well as its mathematical structure and tools, we refer to the respective specialized literature. The following sections focus on a simplified presentation of the atomic model as it results from quantum theory, and the consequences a quantum physical description has for the fundamental processes of absorption and emission of radiation.

7.3.1 The Quantum Mechanical Model of the Atom

The main achievement of the quantum mechanical model of the atom is that it is able to coherently explain the fundamental structure of the atomic emission and absorption spectra. To a certain degree also the Bohr Sommerfeld model was capable of that, but only when based on *Bohr's postulates* that had no physical foundation. In his postulates Bohr effectively assumed certain quantization rules, but he could not derive or explain them.

In the quantum mechanical atom, electrons are not described anymore as particles moving on well-defined trajectories in the Coulomb potential of the atomic core. Instead, they are described by a set of wave functions, each wave function corresponding to a specific state of the electron. These wave functions are the eigenfunctions of the Schroedinger equation for the Coulomb potential and the absolute value of the wave function reflects the probability of finding the electron at a certain location. The respective eigenvalues correspond to a set of quantized values characterizing the respective state of the electron. Some of these values correspond to classical variables such as energy and angular momentum (Fig. 7.9).

As a consequence, each electron state or energy level is unambiguously defined by four quantum numbers:

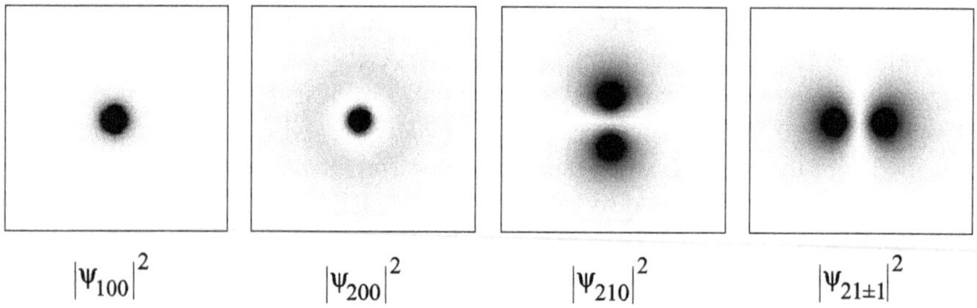

Fig. 7.9 Cross section through the probability density distribution for the electrons of the hydrogen atom, for n = 1, 2; l = 0, 1 and m = 0, ±1

- *Principal quantum number n*:

$$n = 1, 2, 3, \ldots, \infty$$

- *Angular momentum quantum number l*:

$$l = 0, 1, 2, \ldots, (n - 1)$$

- *Magnetic quantum number m*:

$$m = -l, (-l + 1), \ldots, (l - 1), l.$$

- *Spin quantum number s*:

$$s = \pm 1/2$$

The simplest atom, with just one electron, is the *hydrogen atom*. From the quantum mechanical description the energy levels for the hydrogen atom follow as

$$E_{nlms} \equiv E_n = -\frac{m_e e^4}{2(4\pi\varepsilon_0\hbar)^2} \cdot \frac{1}{n^2}. \tag{7.18}$$

Due to the fact that the hydrogen atom has only one electron and thus exhibits perfect spherical symmetry, the energy of the different eigenstates depends only on the principal quantum number, although the corresponding wave functions depend on the other quantum numbers as well (Fig. 7.12). Thus, the electron states with the same principal quantum number n, but different quantum numbers l, m and s are said to be *degenerate*. In this case to each energy level E_n correspond n^2 different wave functions (probability densities), and including the electron spin, $2n^2$ electron states. This is illustrated by the schematic representation of the possible energy levels in Fig. 7.13, e.g. if the degeneracy is lifted by external influences (Fig. 7.10).

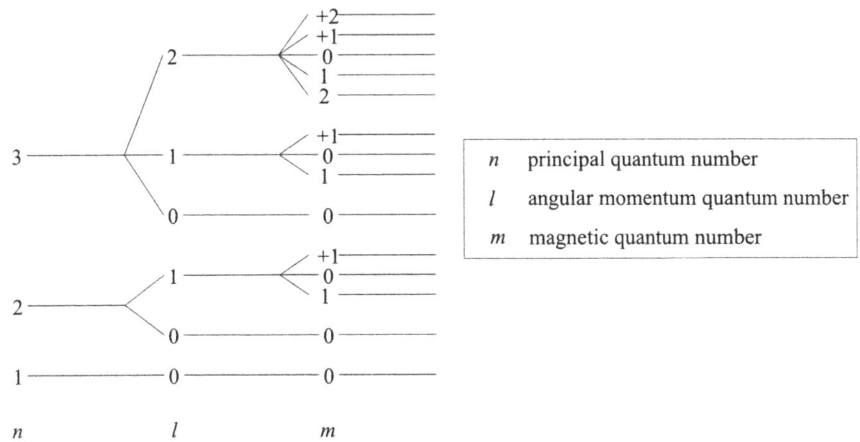

Fig. 7.10 Possible energy levels if the degeneracy is lifted, for the principal quantum numbers n = 1, 2, 3. The electron spin has not been considered in this picture; it leads to an additional splitting of each level in two sublevels

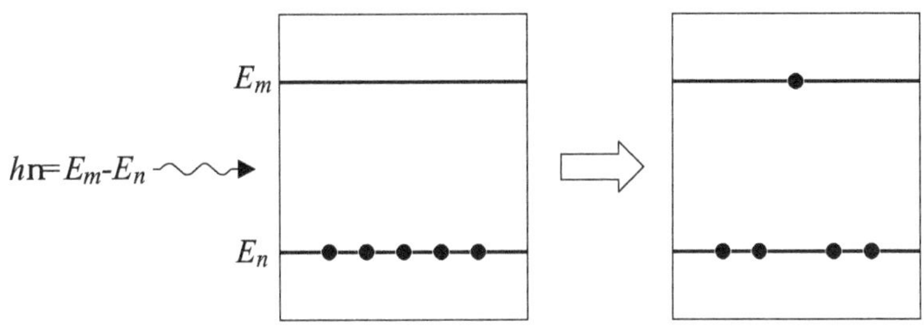

Fig. 7.11 Absorption of a photon

The degeneracy of the energy levels E_n with respect to the angular momentum quantum number l can be lifted by breaking the spherical symmetry of the hydrogen atom's electrical potential, e.g. by an external electrical field. The degeneracy with respect to the magnetic quantum number m and the spin quantum number s can be lifted by an external magnetic field. In this case the energy in the different states also depends on the quantum numbers l, m and s.

In fact, the quantum mechanical description according to the Schroedinger equation reflects only the principal structure of the energy levels of the hydrogen atom. When considering more details, e.g. the spin of the atomic core, then the degeneracy of the energy levels is already lifted without external influence, and the energy values are slightly shifted compared to Eq. 7.57.

For multi-electron atoms the situation is even more complex. The Schroedinger equation then becomes a system of differential equations for the wave functions of each single electron. Since all electrons interact via the Coulomb force and mutually

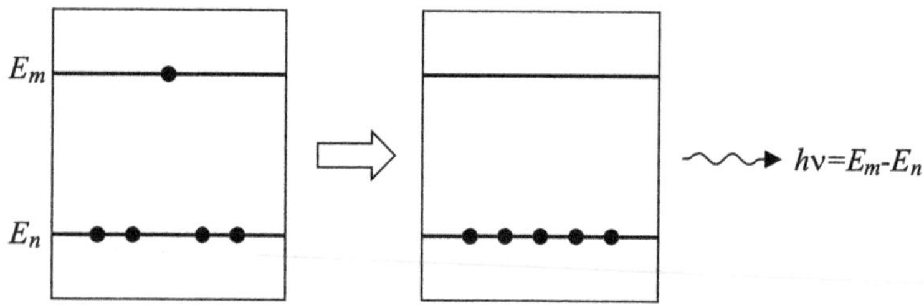

Fig. 7.12 Spontaneous emission of a photon by an excited atom

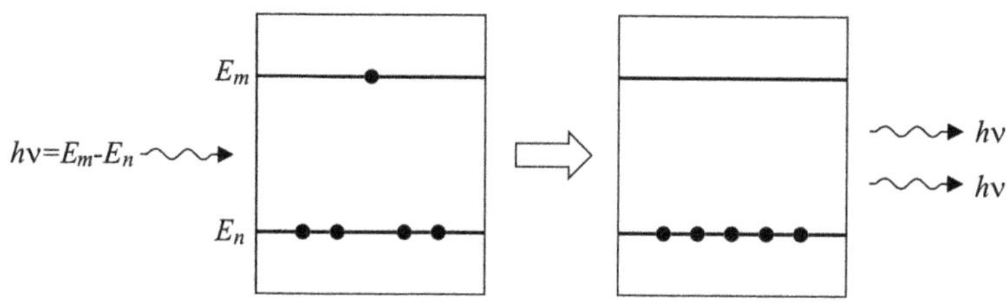

Fig. 7.13 Stimulated emission. The irradiated photon leads to the emission of a second, identical photon

induced magnetic forces, these differential equations are coupled. Effectively, each electron is not subjected to a rotationally symmetric Coulomb potential anymore, but to a potential deformed by all other electrons. This results in additional shifts and splits in the energy levels. In general the energy levels of these atoms can only be calculated using numerical methods.

Two of the three possible fundamental transitions between the atomic energy levels can be derived in the framework of this quantum mechanical model by exposing the atom to an electromagnetic wave as described by Maxwell's equations: The absorption and the stimulated emission. The third fundamental process, the spontaneous emission, can only be derived from a *quantum electrodynamical model* that also subjects the electromagnetic field to a quantum physical description. In the quantum electrodynamical model particles and fields are treated symmetrically: Depending on the measurement, particles can exhibit wave characteristics, and waves or fields can exhibit particle characteristics. In this model, the *photon* is the particle corresponding to electromagnetic fields. In the same way matter consists of particles such as electrons, neutrons and protons, in this model light consists of photons.

7.3.2 Photons

While the Maxwell equations describe many properties of electromagnetic waves very well, they proved unable to explain some other experimental findings.

According to the Maxwell equations the energy of a light wave is proportional to its intensity and does not depend on its frequency. However, already in the late nineteenth century, there were several experimental results known contradictionary to this characteristic:

- Certain chemical reactions can be triggered by light with a sufficiently high frequency, while light at lower frequencies shows no effect, independent of its intensity.
- Electrons can be ejected from the surface of a metal when irradiated with light; however, here again the light needs to have a sufficiently high frequency. Moreover, the kinetic energy of the ejected electrons is proportional to the light frequency and independent of the intensity. This effect is called *photo effect*.

MAX PLANCK was the first to propose a quantization of electromagnetic radiation energy in his explanation of black-body radiation in 1900. In his explanation the radiation consisted of light quantae with an energy proportional to the light's frequency:

$$E = h\nu.$$

The proportionality constant h linking energy and frequency thus is called *Planck's constant*:

$$h = 6.26069 \times 10^{34}\ \text{Js} \quad \text{Planck's constant.}$$

However, Planck interpreted the quantization to be a characteristic of the matter emitting and absorbing the radiation, and not of the radiation field itself. In his explanation of the photo electric effect ALBERT EINSTEIN then showed in 1905 that the quantization was indeed a characteristic of the radiation itself. In his model he generally accepted the validity of the Maxwell equations, but he assumed the energy of the radiation was concentrated in localized quantae of light, the *photons*. This was the beginning of the *photon model* of light.

Based on this model, the following fundamental properties of the photon have been derived and experimentally verified:

- It has no mass, $m_{\text{Ph}} = 0$, and in vacuum always travels at the velocity of light c.
- Its energy correlates to the frequency of the light, and its momentum with the wavelength:

$$E_{\text{Ph}} = h\nu \equiv \hbar\omega, \quad p_{\text{Ph}} = \tfrac{h}{\lambda} \equiv \hbar k \\ \text{with } \hbar = \frac{h}{2\pi} \tag{7.19}$$

E_{Ph} Energy of the photon
p_{Ph} Momentum of the photon

The correlation between photon momentum and wavelength has been proved experimentally by ARTHUR COMPTON for which he received the Nobel prize in 1927.

- The photon is a *boson* and can take on a spin value of +1 or −1. These two spin values correspond to the two possible polarization states of the electromagnetic wave.
- The energy content of a wave is given by the number of photons,

$$\langle W_{em} \rangle = N \cdot E_{Ph} = N \cdot \hbar\omega \tag{7.20}$$

$\langle W_{em} \rangle$ wave energy or averaged field energy
N number of photons

in the same way as the total mass of a piece of material is given by the total mass of the particles it consists of. The energy of the electromagnetic field cannot be divided into smaller portions than the photon energy.

From this it directly follows that the intensity of a light wave is proportional to the photon density:

$$I = c\langle \mathrm{w}_{em} \rangle = c \cdot n \cdot \hbar\omega. \tag{7.21}$$

$\langle W_{em} \rangle$ energy density of the wave
n photon density

7.3.3 Absorption and Emission of Photons

By representing a light wave as a stream of particles, the interaction between eletromagnetic radiation (photons) and matter (atoms) can be explained in very descriptive analogies to classical mechanic processes (Fig. 7.11).

When a photon is absorbed by an atom the photon vanishes. Its energy $E_{Ph} = h\nu$ is transferred to the atom and causes an electron to move to a higher energy level (Fig. 7.14)

Since only specific energy values corresponding to the energy levels E_n of the atom are allowed, only an amount of energy can be transferred that is exactly equivalent to the difference between two energy levels. This is the *energy condition* for the absorption of a photon:

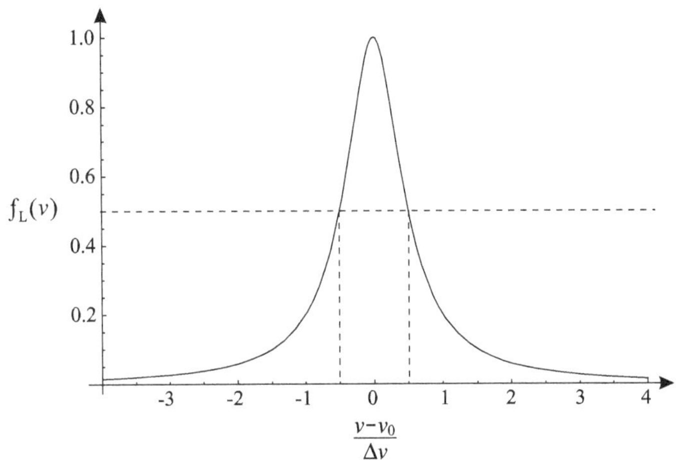

Fig. 7.14 Lorentz distribution. Δν defines the width of the distribution, measured at half of its maximum

$$h\nu = E_m - E_n. \tag{7.22}$$

$h\nu$ energy of the absorbed photon
E_n, E_m energy levels of the atom, with $E_m > E_n$

Photons that do not satisfy this condition for any pair (n, m) of energy levels cannot be absorbed by the atom. When an atom has absorbed one or more photons its electrons do not reside in the configuration with the lowest possible energy anymore: This atom is called *excited* or in an *excited state.*

An excited atom can emit a photon by transitioning into a less excited state. This means one of the electrons transitions to a lower energy level. The energy $E_m - E_n$ released in this process is emitted as a photon with the frequency corresponding to this energy: Thus, in the emission process, a photon is generated.

However, two different emission processes need to be distinguished. In the case of *spontaneous emission* the atom transitions to the lower energy level without external stimulation (Fig. 7.15). This process generates a photon with an energy conforming to the energy condition in Eq. 7.71, but with arbitrary emission direction, polarization and phase.[3] When a large number of spontaneous emission processes occur, then the photons are emitted with statistically uniform distribution of polarization and phase in all directions of space.

In the case of *stimulated emission*, the emission process is triggered by the presence of an electromagnetic wave or, in the picture of photons, by the presence of a photon. The atom again makes the transition to a lower energy state, emitting a photon with the respective energy. In this case, however, the emitted photon is an

[3]The phase of the respective wave function.

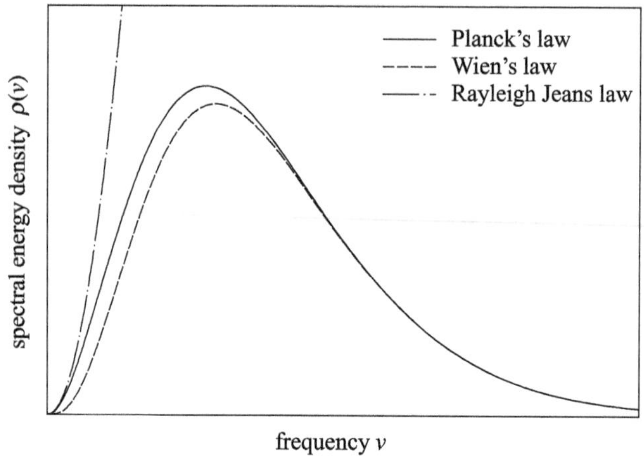

Fig. 7.15 Planck's law and its approximations at low and high frequencies, the Rayleigh-Jeans law and Wien's radiation law

exact duplicate of the triggering photon: Energy, emission direction, polarization and phase are exactly the same (see Fig. 7.16). The triggering photon has to fulfill the energy condition in Eq. 7.71 to enable the stimulated emission process.

Since spontaneous emission does not need a triggering event it can occur at any time as long as the respective atom is in an excited state. This means each atom will remain in the excited state only for a limited period of time. This period of time which can be different for each energy level is called *lifetime* of the respective energy level. The probability per unit of time for a spontaneous emission event leading to a relaxation of the atom from the (higher) energy level E_m to a (lower) energy level E_n is denoted as $w_{mn}^{(Sp)}$. To obtain the total probability for spontaneous emission, the probabilities for transitions to all available lower energy levels have to be added:

$$w_m^{(Sp)} = \sum_{n<m} w_{mn}^{(Sp)}. \tag{7.23}$$

The reciprocal value of the transition probability then is a measure for the average time an atom remains in the excited energy level E_m if only spontaneous emission is considered as a relaxation mechanism. This time is called the lifetime of the energy level E_m.

$$\tau_m = \frac{1}{w_m^{(Sp)}}. \tag{7.24}$$

τ_m lifetime of the mth energy level

If the atom is in its ground state, no spontaneous emission is possible. Thus the lifetime of the lowest energy level is infinite. As a consequence of the *energy-time uncertainty principle*, a limited lifetime implies an uncertainty in the energy of the respective energy level:

$$\Delta E_m = \frac{\hbar}{\tau_m}. \tag{7.25}$$

ΔE_m energy uncertainty of the mth energy level

Due to the uncertainty of the energy, all transitions between the energy levels possess a *linewidth* given by

$$\Delta \nu_{mn} = \frac{1}{h}(\Delta E_m + \Delta E_n) = \frac{1}{2\pi}\left(\frac{1}{\tau_m} + \frac{1}{\tau_n}\right). \tag{7.26}$$

$\Delta \nu_{mn}$ natural linewidth of the transition from E_m to E_n

This broadening of the atomic emission and absorption lines is called natural linewidth, because it is not induced by any external influence, but rather is an inherent characteristic of the respective transition. The *line shape* of a naturally broadened spectral line is given by the *Lorentz distribution* (Fig. 7.17):

$$f_L(\nu) = \frac{\Delta \nu_{mn}/2}{(\nu - \nu_{mn})^2 + (\Delta \nu_{mn}/2)^2}. \tag{7.27}$$

The line shape can be understood when looking at the emission process in the wave model: The amplitude of the emitted wave decays exponentially due to the decreasing population probability during the emission process. The Lorentz shape of the emission line then follows from the Fourier transformation of the emitted wave train.

7.3.4 Einstein's Rate Equations

The transition from a model for emission and absorption processes in individual atoms to a model for the emission and absorption characteristics of a medium consisting of many atoms is done by describing the three fundamental processes by *transition rates*. If the total number of atoms in the observed medium is N and the probability for each atom to be in the energy state E_n is w_n, then statistically an average number of

$$N_n = w_n N \tag{7.28}$$

N_n population number of the energy level E_n

atoms will be in the energy state E_n at any time. Assuming only the energy levels E_1 and E_2 have to be considered, then at all times

$$N = N_1 + N_2 \tag{7.29}$$

holds. If the total number of atoms N does not change over time it is sufficient to describe the population number of one of the levels. The number of photons with a frequency equal to the frequency of the transition between the two energy levels is given by the photon number $\rho(\nu_{21})$.

The atomic absorption processes lead to an increase of the population number of the upper energy level E_2 that is proportional to the photon number $\rho(\nu_{21})$ on one hand and to the population number of the lower energy level E_1 on the other hand. Expressed as a rate of change, this leads to the *transition rate for absorption*:

$$\frac{\mathrm{d}}{\mathrm{d}t}N_2 = -\frac{\mathrm{d}}{\mathrm{d}t}N_1 = B_{12}N_1\rho(\nu_{21}) \tag{7.30}$$

B_{12} Einstein coefficient for absorption

Since N is constant, N_1 always exhibits the reverse behavior of N_2. The proportionality constant in this relation is the so-called Einstein coefficient for absorption and is denominated as B_{12} for historic reasons. Equivalently, the *transition rate for stimulated emission* is given by

$$\frac{\mathrm{d}}{\mathrm{d}t}N_1 = -\frac{\mathrm{d}}{\mathrm{d}t}N_2 = B_{21}N_2\rho(\nu_{21}). \tag{7.31}$$

B_{21} Einstein coefficient for stimulated emission

The spontaneous emission, however, is independent of the photon number, because the presence of a photon is not a prerequisite for this process:

$$\frac{\mathrm{d}}{\mathrm{d}t}N_1 = -\frac{\mathrm{d}}{\mathrm{d}t}N_2 = A_{21}N_2. \tag{7.32}$$

A_{21} Einstein coefficient for spontaneous emission

The transition rates for the three fundamental transition processes combined together result in *Einstein's rate equations* for the population numbers of the energy levels and the photon number:

$$\frac{\mathrm{d}}{\mathrm{d}t}N_1 = -\frac{\mathrm{d}}{\mathrm{d}t}N_2 = \frac{\mathrm{d}}{\mathrm{d}t}\rho(\nu_{21}) = A_{21}N_2 + (B_{21}N_2 - B_{12}N_1)\rho(\nu_{21}), \tag{7.33}$$

or making use of Eq. 7.78:

$$\frac{\mathrm{d}}{\mathrm{d}t}N_2 = -[A_{21} + (B_{21} + B_{12})\rho(\nu_{21})]N_2 + B_{12}\rho(\nu_{21})N \tag{7.34}$$

Einstein's rate equations are valid for equilibrium and nonequilibrium states. However, depending on the observed medium, many more than two energy levels may be relevant, and besides absorption and emission also other, non-radiative excitation and relaxation processes may be present.

7.3.5 *Planck's Law*

First, the rate equations are applied in the description of radiation in the thermal equilibrium or *thermal radiation*. The sun and incandescent lamps are good examples for sources of thermal radiation. A prerequisite for the thermal equilibrium is that the observed body and the surrounding radiation field can be characterized by the same temperature, thus there is no net energy flow between the body and the radiation field: The body emits the same amount of radiation as it absorbs. Thus the population numbers of the energy levels and the photon number are constant:

$$\frac{\mathrm{d}}{\mathrm{d}t}N_1 = \frac{\mathrm{d}}{\mathrm{d}t}N_2 = \frac{\mathrm{d}}{\mathrm{d}t}\rho(\nu) = 0. \tag{7.35}$$

Additionally, from statistical physics it is known that in the thermal equilibrium the population numbers of the energy levels follow the *Boltzmann distribution*:

$$\frac{N_2}{N_1} = \exp\left(-\frac{E_2 - E_1}{k_\mathrm{B}T}\right). \tag{7.36}$$

Introduction of this relation into Einstein's rate equation (Eq. 7.82) results in

$$\rho(\nu) = \frac{A_{21}}{B_{12}} \frac{1}{\exp\left(\frac{h\nu}{k_\mathrm{B}T}\right) - \frac{B_{21}}{B_{12}}} \quad \text{with} \quad E_2 - E_1 = h\nu. \tag{7.37}$$

For $T \to \infty$, also the photon density has to approach infinity, $\rho(\nu) \to \infty$. Thus, it is required that:

$$\frac{B_{21}}{B_{12}} \stackrel{!}{=} \lim_{T\to\infty} \exp\left(\frac{h\nu}{k_\mathrm{B}T}\right) = 1 \quad \Rightarrow \quad B_{12} = B_{21}\,. \tag{7.38}$$

The factor A_{21}/B_{12} in Eq. 7.86 can be determined by comparison with the experimentally well confirmed *Rayleigh-Jeans law*:

$$\rho(\nu) = 8\pi \frac{\nu^2}{c^3} k_B T. \tag{7.39}$$

k_B = 1.38 ×10^{-23} J/K Boltzmann constant
T (absolute) Temperature [K]

The Rayleigh-Jeans law was known since 1900 as a very accurate description of the lower frequency part of thermal radiation.

In the limit of small frequencies the exponential term in Eq. 7.86 can be approximated by a first order series expansion and it follows

$$h\nu \ll k_B T: \quad \rho(\nu) \approx \tfrac{A_{21}}{B_{12}} \tfrac{k_B T}{h\nu} \stackrel{!}{=} 8\pi \tfrac{\nu^2}{c^3} k_B T \quad \Rightarrow \quad A_{21} = 8\pi \tfrac{\nu^2}{c^3} h\nu B_{12} \,. \tag{7.40}$$

With this result Gl. 7.37 becomes

$$\rho(\nu) = 8\pi \frac{\nu^2}{c^3} h\nu \frac{1}{\exp\left(\frac{h\nu}{k_B T}\right) - 1}. \tag{7.41}$$

This is *Planck's law* for the spectral energy density of radiation in a thermal equilibrium (Fig. 7.19).

The spectral energy density $\rho(\nu)$ takes on its maximum at the frequency ν_{max}:

$$\tfrac{d\rho}{d\nu} = 0 \quad \Rightarrow \quad \nu_{max} \approx \tfrac{5}{h} k_B T \,. \tag{7.42}$$

The position of the maximum shifts proportionally with increasing temperature to higher frequencies. This relation is called *Wien's displacement law.* From the position of the inflexion points of the spectral energy distribution its half-width can be calculated:

$$\tfrac{d^2\rho}{d\nu^2} = 0 \quad \Rightarrow \quad \delta\nu \approx \tfrac{5}{h} k_B T = \nu_{max} \,. \tag{7.43}$$

In general, the reciprocal value of the relative linewidth is referred to as the *quality factor Q* or *Q-factor* of a radiating system (see also Sect. 6.2.4):

$$Q = \frac{\nu_{max}}{\delta\nu}. \tag{7.44}$$

Because of Eq. 7.92, for thermal radiation follows Q = 1. This is the lowest possible value for the Q-factor: Purely thermal radiation source is in respect of the Q-factor the exact opposite of the ideal laser with a high-quality optical resonator emitting monochromatic radiation.[4]

[4]With laser resonators a Q-factor of $Q > 10^7$ can be achieved.

The *total energy density* of thermal radiation results from integrating the Planck distribution over all frequencies:

$$\begin{aligned} w_{th} &= \int_0^\infty \rho(\nu)\mathrm{d}\nu \\ &= \tfrac{8\pi}{c^3h^3}(k_\mathrm{B}T)^4 \int_0^\infty \tfrac{x^3}{e^x-1}\mathrm{d}x \quad \text{mit} \quad x = \tfrac{h\nu}{k_\mathrm{B}T} \\ &= \frac{8\pi}{c^3h^3}(k_\mathrm{B}T)^4 \frac{\pi^4}{15}. \end{aligned} \tag{7.45}$$

This means the total energy density of thermal radiation increases with T^4. The intensity of thermal radiation is

$$I_{th} = \frac{1}{2} c w_{th}. \tag{7.46}$$

By the factor ½ it is taken into account that the thermal radiation is emitted in all directions in space, and thus only one half of the radiation is emitted e.g. in the +z direction. From Eqs. 7.94 and 7.95 follows the *Stefan-Boltzmann Law* for the intensity of thermal radiation:

$$I_{th} = \sigma T^4. \tag{7.47}$$

$$\sigma = \frac{4\pi^5}{15}\frac{k_\mathrm{B}^4}{c^2h^3} = 5.67 \times 10^{-8}\frac{W}{m^2K^4} \quad \text{Stefan-Boltzmann constant}$$

A different and self-contained approach to the deduction of Planck's law is based on a statistical description of radiation in a closed cavity. In this case also the relations between the Einstein coefficients for emission and absorption follow directly from the theory.

7.3.6 Population Inversion and Amplification

Einstein's rate equations for a system with two energy levels can be expressed in a more compact form by defining the total number of energy levels N and the *population inversion* D:

$$N = N_1 + N_2, \quad D = N_2 - N_1\,. \tag{7.48}$$

With these definitions the population numbers N_1 and N_2 can written as

$$N_2 = \tfrac{1}{2}(N + D), \quad N_1 = \tfrac{1}{2}(N - D), \tag{7.49}$$

and introducing them into Einstein's rate equation (Eq. 7.82) leads to the following equations for the population inversion and the photon number[5]:

$$\begin{aligned} \tfrac{\mathrm{d}}{\mathrm{d}_t} D &= -(A + 2B\rho)D - AN, \\ \tfrac{\mathrm{d}}{\mathrm{d}_t} \rho &= BD\rho + \tfrac{A}{2}(N + D). \end{aligned} \tag{7.50}$$

For a large photon number the term proportional to A can be omitted in the second equation, resulting in a homogeneous differential equation for ρ:

$$\tfrac{\mathrm{d}}{\mathrm{d}_t} \rho = BD\rho \quad \Rightarrow \quad \rho(t) = \rho(0) \cdot \exp(BD \cdot t). \tag{7.51}$$

From the solution of this simplified equation it is obvious that the photon number only grows over time if $D > 0$: *Only with a positive population inversion a net amplification can be achieved; otherwise the absorption always outhweighs the emission.* In the stationary case follows for the population inversion:

$$\tfrac{\mathrm{d}}{\mathrm{d}_t} D = 0 \quad \Rightarrow \quad D = -\frac{AN}{A + 2B\rho}. \tag{7.52}$$

This means the population inversion in a two-level system is always negative, as long as there are no additional processes present besides the direct transitions between the energy levels E_1 and E_2. At best, for very large photon numbers ($\rho \to \infty$) or vanishing spontaneous emission ($A \to 0$), a population inversion of $D = 0$ is achieved. Consequently, a medium is always absorbing and never amplifying as long as it is in a thermal equilibrium. This suits well to everyday experience. *In order to achieve amplification, population inversion has to be induced in the medium by and additional process.* In case of the laser this process is called *pump process*.

Reference

1. Brehm, J.J. and Mullin, W.J., "Introduction to the Structure of Matter: A Course in Modern Physics," (Wiley, New York, 1989) ISBN 047160531X.

[5] In case of a two level system, the indices of the Einstein coefficients are redundant and thus are omitted in the following.

Chapter 8
The Production of Laser Radiation

In the previous chapter we presented the foundations of the interaction between radiation and material. In the end, the necessary criterion to produce laser radiation was derived: population inversion. Population inversion is the prerequisite for the predominance of stimulated emission, the basic process of the laser.

8.1 The Laser Principle

The laser converts entropy.[1] It transforms energy of low quality into energy of the highest quality: *laser energy is energy of the highest quality*. A characteristic of this is the high spatial and temporal energy density that can be attained in the focus of laser radiation. The initial form of energy used to pump lasers is generally, on the contrary, less concentrated: this energy, for example, consists of thermal light from pump lamps or electrical current. This energy is transported into the laser medium and stored there as *population inversion*. The energy that is buffered and bundled is finally transformed into laser radiation via *stimulated emission*. The characteristics of this transformation process, together with the feedback effect of the laser resonator, lead to the high quality of the transmitted energy: the energy transferred by the individual atoms is effectively synchronized; laser radiation represents, thus, energy with a high degree of order (Fig. 8.1).

After this illustrative overview of the internal operating mode of the laser, the individual process steps will be investigated in more detail in the rest of this chapter.

[1]Entropy can be seen as a measurement for the degree of order or the quality of energy. Lower entropy means a high amount of order and high quality. Irreversible processes always lead to an increase of entropy.

R. Poprawe et al., *Tailored Light 1*, RWTHedition,
DOI 10.1007/978-3-642-01234-1_8

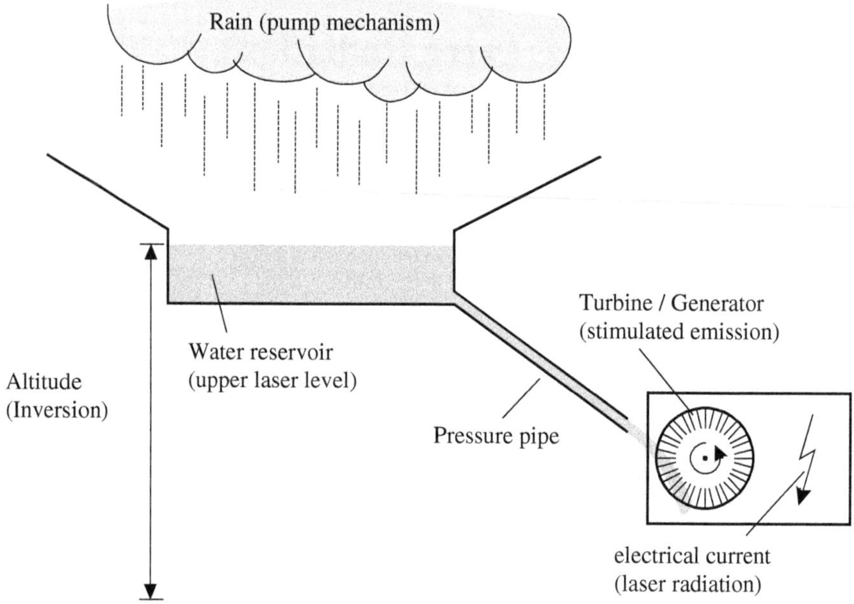

Fig. 8.1 Generation of electrical current through hydroelectric power as an analogy to the energy transformation in the laser medium. The laser medium is replaced by the water circulation. Rain fills the catchment basin, and via the pressure pipe the stored water is led to the turbine in concentrated form and transformed into electrical energy

8.2 Producing Population Inversion

As the argumentation in Sect. 7.4.9 showed, the production of the population inversion is accorded a key role: without inversion, amplification and, thus, laser emission is not possible.

Initial point for the production of population inversion is the laser medium in a thermal equilibrium. The energy levels are occupied according to the Boltzmann distribution:

$$N_n = N_0 \exp\left(-\frac{E_n - E_0}{k_B T}\right). \tag{8.1}$$

E_0, N_0 Energy and population number of the ground state
E_n, N_n Energy and population number of the nth energy state
T Absolute temperature
k_B Boltzmann constant.

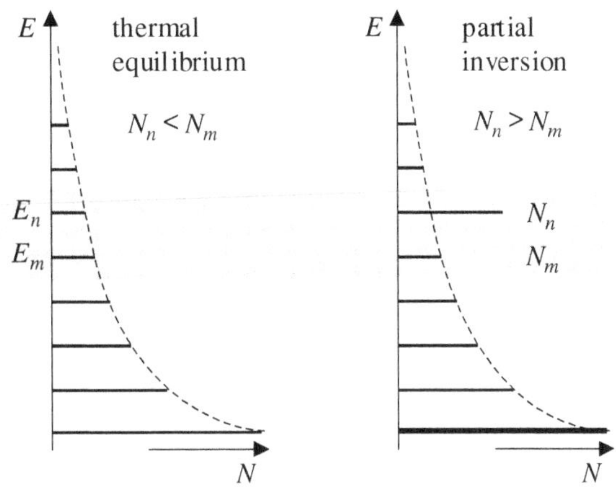

Fig. 8.2 *Left* thermal occupation of the energy levels according to the Boltzmann statistic; *right* partial inversion through increased population density in *n*th level

In order to produce amplification for a transition between two energy levels E_n and E_m, *partial inversion* has to be reached with reference to these two levels, which means the following

$$N_n > N_m \quad \text{for} \quad E_n > E_m \tag{8.2}$$

has to occur (Fig. 8.2). The inversion condition can, as shown in Sect. 7.4.9, only be produced through an additional *pump process*, which excites atoms from lower energy levels into the upper laser level E_n. The direct transition from the lower into the upper laser level is inappropriate to pump lasers, since an inversion cannot be reached in this way, but rather only an equal occupation of the levels. *The two-level system* is not capable of operating a laser.

8.2.1 Three-Level Systems

For the three-level system the pump process is, therefore, carried out using an additional energy level, the *pump level*. The pump level has a higher energy than the laser level. By means of the pump mechanism, atoms of the lower laser level, which is simultaneously the ground state of the laser medium, are excited into the pump level. From there the atoms pass over into the upper laser level via a quick, spontaneous process, and are then at the disposal of the laser transition (Fig. 8.3).

Prerequisite for the inversion in the three-level laser is that

- The pump rate is large,
- The transition from the pump level into the upper laser level occurs very quickly, and

Fig. 8.3 Three-level system

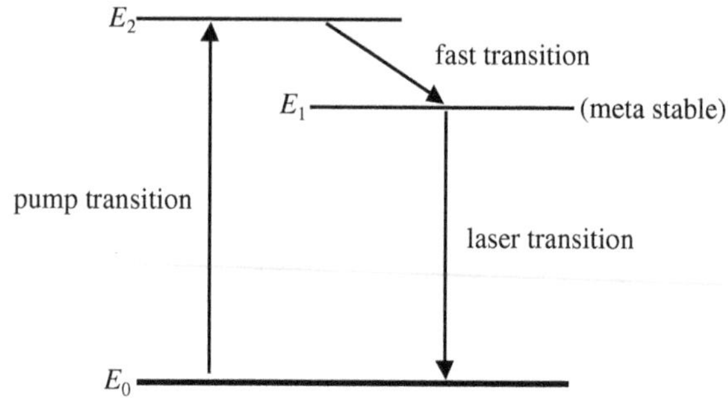

- The upper laser level is *metastable*, which means it exhibits a long life time and thus the laser transition from E_1 to E_0 occurs fewer times than the transition from E_2 to E_1:

$$\tau_2 \ll \tau_1. \tag{8.3}$$

τ_1, τ_2 Life time of the energy states E_1 and E_2.

This process assures that atoms can collect in the upper level, on the one hand, thus producing inversion, and that the pump level is only weakly filled at high pump rates, on the other, whereas the effectiveness of the pump mechanism is preserved.

8.2.2 Four-Level Systems

A further approach to increase inversion is provided by the *four-level system*. In contrast to the three-level laser, the lower laser level is no longer identical with the ground state (Fig. 8.4).

A short life time in the lower laser level leads to quick relaxation into the ground state which has a lower energy. The lower occupation of the lower laser level

Fig. 8.4 The four-level system

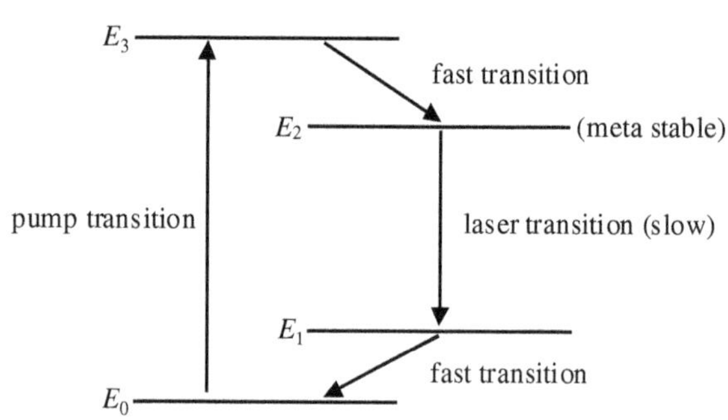

contributes to the inversion, and the quick filling of the ground state makes the pump mechanism more effective. In contrast to the three-level system, significantly lower pump power is needed for inversion.

Two essential factors contribute to a high population inversion in the case of the four-level laser:

- A *high pump rate* leads to a higher occupation of the upper laser level and increases thereby the inversion, and
- *Cooling* reduces the occupation of the lower laser level and, therefore, increases the population inversion as well. A lower temperature leads to a reduction of the equal balance occupation of the higher energy state according to the Boltzmann distribution, Eq. 8.1 (Fig. 8.5).

The efficiency of the total laser process,

$$\eta = \frac{P_{\text{out}}}{P_{\text{in}}} = \frac{E_{\text{out}}}{E_{\text{in}}}, \tag{8.4}$$

P_{out}, E_{out} Laser power output, or output energy
P_{in}, E_{in} Input power, or input energy

is proportional to the *quantum efficiency*

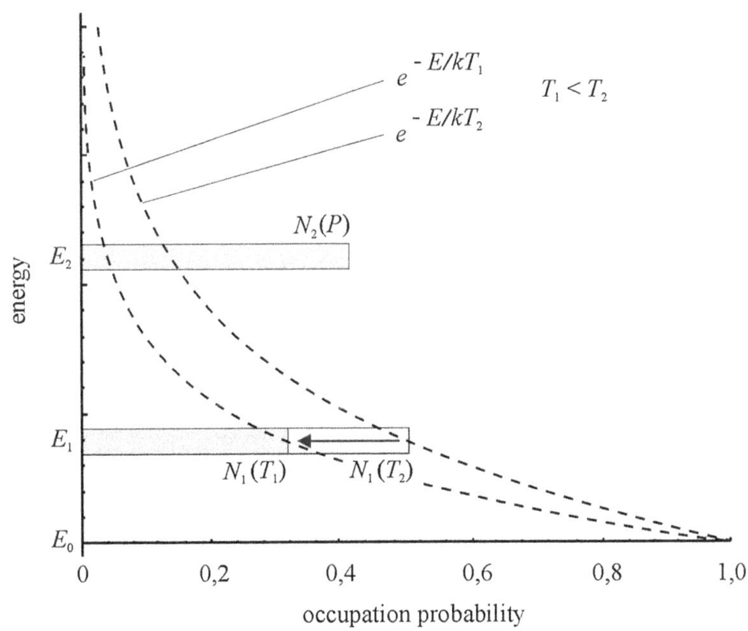

Fig. 8.5 Cooling reduces the population of the lower laser level and thus facilitates the occurrence of inversion

$$\eta_Q = \frac{\nu_{\text{Laser}}}{\nu_{\text{Pump}}} = \frac{E_2 - E_1}{E_3 - E_0}, \tag{8.5}$$

since the pump mechanism has to expend the energy difference $E_3 - E_0$ to excite each atom, whereas the laser transition only has to transform the energy $E_2 - E_1$ into radiation. In order to reach a higher total efficiency, the energy of the pump level may not be much higher than the laser level.

8.2.3 *Pump Mechanisms*

The details of the pump mechanisms used in the individual lasers are explained in connection with the laser systems in Chap. 9. An overview of the most important mechanisms, however, will be given here along with their functional principles:

- *Optical Pumping*: the laser medium is excited optically by light from flash lamps, arc lamps, or lasers with an appropriate frequency. For this, the frequency of the pump light has to be larger than the laser transition. This pump mechanism is only suitable for optically dense laser media, since the pump light has to be absorbed effectively. Therefore, *solid-state lasers* are primarily pumped optically. Pumping solid-state lasers with *diode lasers* should be highlighted here as a forward-looking technology.
- Pumping through *Gas Discharge*: the atoms of the *gaseous* laser medium are excited through collisions with high energy electrons or other atoms. To produce the gas discharge, high-frequency modulated high voltage is used (*HF excitation*).
- Pumping through *Current Injection*: In the case of the semiconductor laser, the current injection leads to population inversion between the *conduction* and *valence band*. In a so-called *pn transition*, the optical relaxation occurs through the laser transition. Due to the direct transition of the electrical input power in optical output power, semiconductor lasers exhibit the highest efficiency, at over 50 %.
- *Chemical Pumping*: the gaseous laser medium is excited by strongly exothermic,[2] chemical reactions. The *excimer laser* takes its name from the molecular substances mostly used, so-called *excimers*. What is remarkable is that excimer lasers function with a two-level system in principle, although the normal energy level is unstable. The excimers which are not excited decay quickly into their individual chemical components, so that the normal energy level is effectively always empty.

[2]Exothermic reaction: reaction during energy excess, i.e., the reaction releases energy.

- *Thermodynamic Pumping*: suitable for gas lasers, this mechanism uses *nonadiabatic processes*,[3] such as a quick heating of the laser gas to produce inversion. The corresponding lasers are called gas-dynamic lasers.

8.3 The Rate Equations of the Laser

In this section, the inner processes in the laser medium are described in more detail on the basis of rate equations. The four-level system is chosen as a starting point for this. Figure 8.6 illustrates the energy diagram with the names of the transitions involved.

The atoms are excited by *external pumping* at the pump rate P^* from the normal level into the upper pump level E_3. From there the atoms pass over into the upper laser level through spontaneous transitions at the *internal pump rate* P. Between the laser levels E_2 and E_1, laser transition by stimulated emission takes place on the one hand, and, on the other, spontaneous emission and absorption also occur. The appropriate rates were established in the last chapter. Added to this is the total rate for spontaneous emission from the upper laser level in other levels as E_1,

$$A_{2*} = \left(\sum_n A_{2n} \right) - A_{21}. \tag{8.6}$$

For this, what must be considered is that the spontaneous emission occurs in all lower energy levels. Lastly, atoms fall back into the ground state, also through spontaneous processes. The coefficients A_{10} summarize all spontaneous transitions from the lower laser level in the ground state: these transitions can also proceed indirectly via further intermediate levels.

In the rate equations only both of the laser levels are included. Hence, only the internal pump rate P appears there and not the external pump rate P^*. Both of the pump rates are connected to each other by the *internal pump efficiency* η_P

$$\eta_P = \frac{P}{P^*}. \tag{8.7}$$

Here the assumption is made that the population of the lowest energy level N_0 is always much larger than that of the laser level, and, therefore, remains nearly constant. Then a constant pump rate P, independent of the laser process, can also be assumed.

Based on Sect. 7.4.6, the rate equations for the population numbers of the laser level read as follows:

---

[3] Adiabatic process: the process runs so slowly that the gas is always found in a momentary equilibrium.

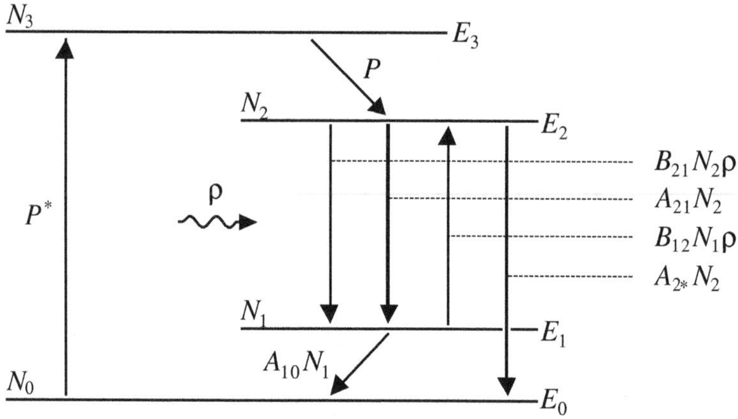

Fig. 8.6 Energy levels and transitions of the four-level laser

$$\begin{aligned}\frac{\mathrm{d}}{\mathrm{d}t}N_2 &= P + B_{12}\rho N_1 - B_{21}\rho N_2 - A_{21}N_2 - A_{2*}N_2\\ \frac{\mathrm{d}}{\mathrm{d}t}N_1 &= B_{21}\rho N_2 - B_{12}\rho N_1 + A_{21}N_2 - A_{10}N_1.\end{aligned} \tag{8.8}$$

The rate equations for the photon number ρ is also taken over from the description of the two-level system and modified slightly:

$$\frac{\mathrm{d}}{\mathrm{d}t}\rho = B_{21}N_2\rho - B_{12}N_1\rho + FA_{21}N_2 - \beta\rho + K. \tag{8.9}$$

F Resonator selectivity
β Resonator loss factor
K Photon injection rate

The *injection rate K* describes an additional radiation of photons in the resonator by an external source. In this way, for example, the operation as an amplifier can be treated; K is then the input signal.

The *resonator selectivity F* accounts for the spontaneous emission releasing photons with any propagation direction. The resonator "traps" only a part of the photons that are emitted in a specific solid angle range. This solid angle range is given by the divergence angle of the beam in the resonator, which minimally corresponds to the diffraction angle,

$$\theta_B = \frac{\lambda}{\pi w_0} \tag{8.10}$$

λ Wavelength of the laser radiation
w_0 Waist radius of resonator mode

(cf. Sect. 5.2.1). The resonator selectivity can be estimated as the relationship of the solid angle covered by the resonator to the complete solid angle 4π:

$$F = \frac{1}{4\pi}\int_0^{2\pi}\int_{-\theta_B}^{\theta_B} \sin\theta \mathrm{d}\theta \mathrm{d}\phi \approx \frac{1}{4\pi} 2\pi \int_{-\theta_B}^{\theta_B} \theta \mathrm{d}\theta = \frac{1}{2}\theta_B^2 = \frac{\lambda^2}{2\pi^2 w_0^2}. \tag{8.11}$$

Typical values for the resonator selectivity are $F = 10^{-6}$–10^{-4}.

The *resonator loss factor* β indicates the outcoupling of radiation by the resonator mirror:

$$\beta = (1 - R_1 R_2)\frac{c}{2L}. \tag{8.12}$$

R_1, R_2 Reflectivity of the first and second resonator mirror
L Resonator length.

For every reflection at a resonator mirror, only the part of the radiation given by the reflectivity remains in the system. Division by the roundtrip period of the resonator $2L/c$ results in the loss factor.

The rate equations Eqs. 8.8 and 8.9 can only be solved with difficulty in this general form. It is, however, obvious that population inversion can only be attained when the population density of the lower laser level E_1 is decreased much more quickly than that of the upper laser level E_2. This means that the transition coefficient A_{10} has to be very large:

$$A_{10} \gg A_{21}, B_{21}\rho. \tag{8.13}$$

Then the term proportional to A_{10} in the rate equation for the lower laser level, and the remaining terms can be neglected:

$$\frac{\mathrm{d}}{\mathrm{d}t}N_1 = -A_{10}N_1 \Rightarrow N_1 \sim \mathrm{e}^{-A_{10}t} \to 0, \tag{8.14}$$

which means the population of the lower laser level quickly approaches zero. For this reason the population inversion D can be replaced by the population density of the upper laser level,

$$D = N_2 - N_1 \approx N_2, \tag{8.15}$$

and there are then only two more rate equations to be solved:

$$\begin{aligned}\frac{\mathrm{d}}{\mathrm{d}t}D &= P - (B_{21}\rho + A_{21} + A_{2*})D \\ \frac{\mathrm{d}}{\mathrm{d}t}\rho &= B_{21}D\rho + FA_{21}D - \beta\rho + K.\end{aligned} \tag{8.16}$$

For the conversion of Eqs. 8.8 and 8.9 into Eq. 8.16, the basic relationship $B_{12} = B_{21}$ was used. To simplify the following discussion, the normalizations

$$\rho = \frac{A_{21}}{B_{21}}\rho', \quad D = \frac{A_{21}}{B_{21}}D', \quad P = \beta\frac{A_{21}}{B_{21}}P', \quad K = \beta\frac{A_{21}}{B_{21}}K' \tag{8.17}$$

are introduced. The time scale is also normalized,

$$t = \frac{1}{A_{21}}\tau \quad \Rightarrow \quad \frac{\mathrm{d}}{\mathrm{d}t} = A_{21}\frac{\mathrm{d}}{\mathrm{d}\tau}, \tag{8.18}$$

so that τ indicates the time relative to the spontaneous lifetime of the upper laser level $1/A_{21}$. In the normalized form without parameters the rate equations now read as follows

$$\begin{aligned}\frac{\mathrm{d}}{\mathrm{d}\tau}D' &= -(1 + \delta + \rho')D' + \alpha P' \\ \frac{\mathrm{d}}{\mathrm{d}\tau}\rho' &= D'\rho' + FD' - \alpha\rho' + \alpha K'\end{aligned} \tag{8.19}$$

with $\alpha = \frac{\beta}{A_{21}}, \quad \delta = \frac{A_{2*}}{A_{21}}$.

The rate equations are one system of two coupled differential equations. The nonlinear coupling over the product of D' and ρ' complicates their exact solution.

8.3.1 Solving the Rate Equations for Stationary Operation

For a large pulse duration or in *cw operation*,[4] a dynamic equilibrium between the running processes is reached in the laser medium. Then, the population inversion and the photon density no longer, which means the rate equations in Eq. 8.19 can be solved statically:

$$\frac{\mathrm{d}}{\mathrm{d}\tau}D' = 0, \quad \frac{\mathrm{d}}{\mathrm{d}\tau}\rho' = 0. \tag{8.20}$$

For the normalized population inversion, what immediately follows is

[4] *cw* = continuous wave, not pulsed.

$$D' = \frac{\alpha P'}{1 + \delta + \rho'}. \tag{8.21}$$

Inserted in the rate equation for the normalized photon density, this results in a quadratic equation with the solution

$$\rho' = \frac{1}{2}\left[P' + K' - 1 - \delta + \sqrt{(P' + K' - 1 - \delta)^2 + 4(FP' + (1+\delta)K')}\right]. \tag{8.22}$$

The second solution of the quadratic equation, with the minus sign in front of the root term, is eliminated, since the photon density has to be positive.

Dependent upon the parameters P' and K', the behavior of the solutions in Eqs. 8.21 and 8.22 will be discussed in three different ranges.

Unamplified Signal Transmission

In this case, the medium is not pumped, but an input signal is coupled in:

$$P' = 0, \quad K' > 0. \tag{8.23}$$

Inserted into Eqs. 8.21 and 8.22, what follows for the normalized form is

$$D' = 0, \quad \rho' = \frac{1}{2}\left[K' - 1 - \delta + \sqrt{(K' - (1+\delta))^2 + 4(1+\delta)K'}\right] = K', \tag{8.24}$$

and after reverse substitution corresponding to Eq. 8.17, the inversion density and the photon density result

$$D = 0, \quad \rho = \frac{K}{\beta}. \tag{8.25}$$

The inversion is minimal in this case, which means zero. Due to the approximation added to Eq. 8.15, $D = N_2$, the inversion cannot become negative. This means that the signal losses are neglected on the grounds of the absorption in the medium. The photon density ρ depends, therefore, only on the input signal K and the resonator losses β: in the unpumped case, an equilibrium is reached between the signal rate and the resonator losses.

Weak Pumping: Spontaneous Emission

If the medium is now pumped weakly and a signal is no longer coupled in,

$$P' \ll 1 + \delta, \quad K' = 0, \tag{8.26}$$

then the normalized photon density results in

$$\begin{aligned}\rho' &= \frac{1}{2}\left[P' - (1+\delta) + \sqrt{(P' - (1+\delta))^2 + 4FP'}\right] \\ &= \frac{1}{2}\left[P' - (1+\delta) + |P' - (1+\delta)|\sqrt{1 + \frac{4FP'}{(P' - (1+\delta))^2}}\right].\end{aligned} \tag{8.27}$$

Due to the condition Eq. 8.26, the value can be solved to a negative sign:

$$P' < 1+\delta \quad \Rightarrow \quad |P' - (1+\delta)| = -[P' - (1+\delta)]. \tag{8.28}$$

Series expansion of the root leads to

$$\begin{aligned}\rho' &\approx \frac{1}{2}\left[P' - (1+\delta) - (P' - (1+\delta))\left(1 + \frac{2FP'}{(P' - (1+\delta))^2}\right)\right] \\ &= \frac{FP'}{1+\delta-P'} \\ &\approx \frac{F}{1+\delta}P' \ll 1.\end{aligned} \tag{8.29}$$

Inserting the previously determined, normalized photon density in Eq. 8.21 results in

$$D' = \frac{\alpha P'}{1+\delta+\rho'} \approx \frac{\alpha}{1+\delta}P'. \tag{8.30}$$

for the normalized population inversion. In this way, the unnormalized values again follow by means of back substitution:

$$\rho \approx \frac{FP}{(1+\delta)\beta}, \quad D \approx \frac{P}{A_{21}(1+\delta)} = \frac{P}{A_{21} + A_{2*}}. \tag{8.31}$$

Photon density as well as inversion density increase linearly with the pump rate P. The inversion results from the equilibrium between the pump rate and the *total spontaneous emission* from the upper laser level. The photon density is determined by the resonator losses β and the resonator selectivity F. The dependency of the selectivity also shows that the spontaneous emission is the dominant process: since the spontaneous emission occurs uniformly in all spatial directions, only the part proportional to F is caught by the resonator from the total number of photons and fed to the laser process once again. For good resonators F is very small (10^{-6}–10^{-4}); therefore, the photon density along with the pump rate increases in this region only very slowly.

Strong Pumping: Laser Operation
For limiting cases, very large pump rates,

$$P' \gg 1+\delta, \quad K' = 0, \tag{8.32}$$

can be assumed directly from Eq. 8.27 for the normalized photon density. In this case, the value retains, however, the negative sign. Once again, the root can be expressed by a series extension in the same way, such that this time

$$\begin{aligned}\rho' &\approx \frac{1}{2}\left[P' - (1+\delta) + (P' - (1+\delta))\left(1 + \frac{2FP'}{(P' - (1+\delta))^2}\right)\right]\\ &= P' - (1+\delta) + F\frac{P'}{P' - (1+\delta)} \approx P'\end{aligned} \tag{8.33}$$

holds. From this the normalized population inversion results,

$$D' = \frac{\alpha P'}{1+\delta+P'} \approx \frac{\alpha P'}{P'} = \alpha, \tag{8.34}$$

and subsequently the unnormalized values as well,

$$\rho \approx \frac{P}{\beta}, \quad D \approx \frac{A_{21}}{B_{21}}\alpha = \frac{\beta}{B_{21}}. \tag{8.35}$$

For large pump rates *saturation inversion* is reached, which means that the inversion no longer increases along with increasing pump rates (Fig. 8.7). The saturation inversion is given by the equilibrium between the resonator losses and the *stimulated emission*; the spontaneous emission is no longer significant for large pump rates.

The photon density continues to increase proportionally to the pump rate. In this region, it is dependent solely on the resonator losses; the resonator selectivity F no longer has an effect due to the subordinate role of the spontaneous emission. This

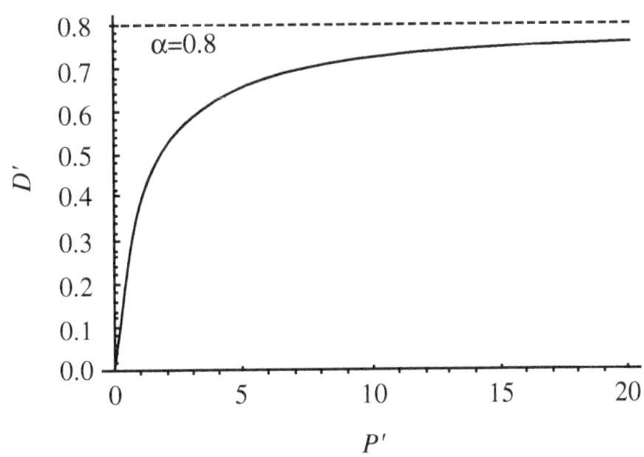

Fig. 8.7 Normalized inversion density D' over the normalized pump rate P'. For large pump rates, D' approaches the saturation value α asymptotically

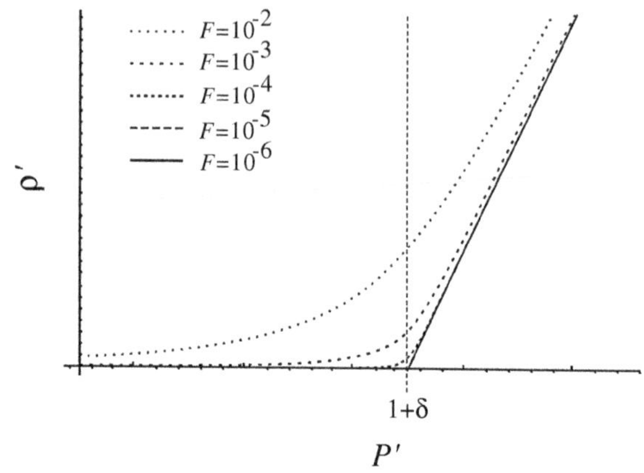

Fig. 8.8 The normalized photon density ρ' over the normalized pump rate P', plotted for different resonator selectivities F. The stronger the resonator properties are, i.e., the smaller F is, the more distinct the transition in laser operation is

results in the photon density increasing significantly more steeply with the pump rate (Fig. 8.8). This region is where *laser operation* takes place.

8.3.2 The Laser Threshold

The limiting cases treated in the last section of small and large pump rates have shown that lasers under a specific threshold value—the *laser threshold*—behave as a thermal light source does: the radiation primarily results from spontaneous emission. Above the threshold in the *laser region*, however, stimulated emission dominates.

The laser threshold lies at (Fig. 8.8).

$$P'_{\text{Threshold}} = 1 + \delta \quad \Rightarrow \quad P_{\text{Threshold}} = \beta \frac{A_{21} + A_{2*}}{B_{21}}. \tag{8.36}$$

The sum of the coefficients A_{21} and A_{2*} represents collected spontaneous transitions from the upper laser level. The laser threshold is determined, therefore, by the relationship of the total losses—resonator losses and losses through spontaneous emission processes—to the stimulated emission. If one inserts the photon density according to Eq. 8.35 in the threshold condition Eq. 8.36, the condition follows

$$\rho_{\text{Threshold}} = \frac{P_{\text{Threshold}}}{\beta} \quad \Rightarrow \quad B_{21}\rho_{\text{Threshold}} = A_{21} + A_{2*} \quad \Leftrightarrow \quad w_{\text{stim}} = w_{\text{spon}}, \tag{8.37}$$

w_{stim} Transition probability for stimulated emission
w_{spon} Total transition probability for spontaneous emission

which means that at the threshold the transition probabilities for stimulated and spontaneous emission are the same. Above the threshold, the internal radiation field has then grown so much that the probability for stimulated emission is much larger than for spontaneous emission. The atoms are forced to pass over into the lower laser level by stimulated emission before the transition can occur by spontaneous emission.

The threshold, however, only exists as long as the system is enclosed in an effective resonator. This is expressed by the dependency of the curves in Fig. 8.8 on the resonator selectivity F. The more F approaches one, the less the spontaneous emission is suppressed by the resonator. In the limiting case $F = 1$, a resonator is not present and the system emits purely thermally as a cavity radiator. The transition into the laser operation does not take place.

In a thermodynamic sense, the transition from the thermal emission to laser emission at the laser threshold represents a *phase transition of the second kind*: the system passes over into a state of a higher order discontinuously.[5] This new state is characterized by *maximum coherence* of the radiation. As a consequence, laser radiation exhibits

- The highest energy flow density,
- The highest monochromatism, corresponding to the smallest line width, and
- Minimal amplitude fluctuations.

Other well-known phase transitions of the second kind are the transition from paramagnetism to ferromagnetism at the Curie temperature or the transition from metallic conduction to superconduction at the critical temperature. In these cases, a discontinuous transition also takes place into a higher state of order. In the thermodynamic sense, a significant difference is, however, that both of the last transitions are observed in systems in thermodynamic equilibrium, whereas the laser is not found in equilibrium: therefore, this kind of phase transition is called a *nonequilibrium phase transition*.

8.3.3 Amplification

This section aims to obtain expressions for the amplification process of the intensity in the laser medium as well as the absolute signal amplification after passage of an amplifying medium of the length L.

[5]Phase transition of the first kind: energy or entropy of the system as a function of the order parameters change discontinuously, i.e., they make a jump. Phase transition of the second kind: energy or entropy change continuously. The first derivation is, however, discontinuous, i.e., the function has a "bend." In the case of the laser, the pump output is the order parameter.

Section 7.4.9 has already shown that the population inversion determines the amplification. Thus, an equation for the inversion D is derived once more. As a starting point, the rate equations, Eq. 8.8, are selected for the population figures of the laser level, since the population of the lower laser level should not be neglected this time. In the stationary case, the equations read as follows,

$$\begin{aligned} P + B_{21}\rho(N_1 - N_2) - (A_{21} + A_{2*})N_2 &= 0 \\ B_{21}\rho(N_2 - N_1) + A_{21}N_2 - A_{10}N_1 &= 0. \end{aligned} \tag{8.38}$$

These equations are solved by

$$\begin{aligned} N_1 &= \frac{B_{21}\rho + A_{21}}{B_{21}\rho(A_{10} + A_{2*}) + A_{10}(A_{21} + A_{2*})} P \\ N_2 &= \frac{B_{21}\rho + A_{10}}{B_{21}\rho(A_{10} + A_{2*}) + A_{10}(A_{21} + A_{2*})} P, \end{aligned} \tag{8.39}$$

which can also be given in the form of the population inversion as a function of the photon density:

$$D \equiv N_2 - N_1 = \left(\frac{A_{10} - A_{21}}{A_{10}(A_{21} + A_{2*})}\right) \cdot \frac{P}{1 + \frac{A_{10} + A_{2*}}{A_{10}(A_{21} + A_{2*})} B_{21}\rho}. \tag{8.40}$$

By inserting the *small-signal inversion* D_0 and the *effective lifetime of the signal amplification* τ_{eff}

$$D_0 = \frac{A_{10} - A_{21}}{A_{10}(A_{21} + A_{2*})} P, \quad \frac{1}{\tau_{eff}} = \frac{A_{10}(A_{21} + A_{2*})}{A_{10} + A_{2*}}, \tag{8.41}$$

one obtains the expression for the *saturable inversion*:

$$D(\rho) = \frac{D_0}{1 + B_{21}\tau_{\text{eff}}\rho}. \tag{8.42}$$

This result can be inserted into the rate equation for the photon density, Eq. 8.9, whereas the spontaneous emission is neglected in the resonator modes as are the resonator losses and the input signal:

$$\frac{\mathrm{d}}{\mathrm{d}t}\rho = B_{21}D(\rho)\rho = \frac{B_{21}D_0}{1 + B_{21}\tau_{\text{eff}}\rho} \cdot \rho. \tag{8.43}$$

Now, the photon density can be converted to the intensity and the temporal dependency to the dependency of the coordinates in the propagation direction,

$$I = c\langle w_{em}\rangle = c \cdot \hbar\omega \cdot \rho, \quad z = ct \Rightarrow \frac{\mathrm{d}}{\mathrm{d}t} = c\frac{\mathrm{d}}{\mathrm{d}z}, \tag{8.44}$$

so that the equation for the *course of the intensity in the laser medium* finally follows:

$$\frac{\mathrm{d}}{\mathrm{d}z}I(z)=\frac{g_0}{1+I(z)/I_S}\cdot I(z)\quad\text{with}\quad g_0=\frac{B_{21}D_0}{c},\ I_S=\frac{\hbar\omega c}{B_{21}\tau_{\mathrm{eff}}}. \tag{8.45}$$

g_0 Small-signal amplification coefficient
I_S Saturation intensity.

Small-signal Amplification
For very small intensities $I \ll I_S$, the input signal increases exponentially corresponding to the small-signal amplification coefficient:

$$I \ll I_S:\quad \frac{\mathrm{d}I}{\mathrm{d}z}=g_0 I \quad\Rightarrow\quad I(z)=I(0)\cdot\exp\{g_0 z\}. \tag{8.46}$$

The small-signal amplification coefficient g_0 is proportional to the population inversion D_0 and thus to the pump rate P (cf. Eqs. 8.41 and 8.45). The *small-signal amplification* for a simple signal pass through the laser medium is defined then as the relationship of the output signal to the input signal:

$$G_0=\frac{I(L)}{I(0)}=\exp\{g_0 L\}. \tag{8.47}$$

L Length of the amplifying medium
G_0 Small-signal amplification.

Nonlinear Amplification
If the intensity becomes comparable to the saturation intensity, then the amplification becomes *nonlinear*, which means the amplification coefficient is dependent upon the signal amplitude: $G = G(I)$.

In the nonlinear amplification region, an approximation cannot be used for Eq. 8.45. The direct solution of this differential equation is difficult; it can, however, be converted to

$$\left(\frac{1}{I(z)}+\frac{1}{I_S}\right)\mathrm{d}I=g_0\mathrm{d}z, \tag{8.48}$$

and then both sides can be integrated with given input and output intensities $I(0)$ and $I(L)$:

$$\int_{I(0)}^{I(L)} \left(\frac{1}{I} + \frac{1}{I_S}\right) \mathrm{d}I = \int_{0}^{L} g_0 \mathrm{d}z \quad \Rightarrow \quad \ln\left(\frac{I(L)}{I(0)}\right) + \frac{I(L) - I(0)}{I_S} = g_0 L \equiv \ln G_0. \tag{8.49}$$

In this way the nonlinear amplification can be given in the following form after passing through a laser medium of the length L:

$$G \equiv \frac{I(L)}{I(0)} = G_0 \exp\left\{-\frac{I(L) - I(0)}{I_S}\right\}. \tag{8.50}$$

G Nonlinear amplification.

An explicit expression cannot, however, be given for the output intensity as a function of the input intensity.

Saturation
For very large intensities $I \gg I_S$, the amplification reaches saturation and the following

$$I \gg I_S : \frac{g_0}{1 + I/I_S} \approx \frac{g_0}{I/I_S} \quad \Rightarrow \quad \frac{\mathrm{d}I}{\mathrm{d}z} = g_0 I_S = \text{const.} \tag{8.51}$$

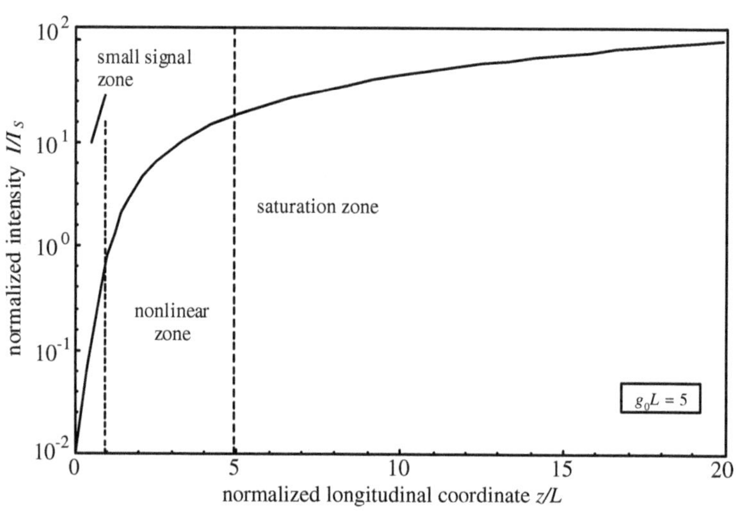

Fig. 8.9 Intensity plotted logarithmically as a function of the propagation length in the active medium, and the three amplification regions

results from Eq. 8.45. In the saturation range, the intensity thus increases linearly along the path covered in the laser medium:

$$I(L) = I(0) + g_0 L I_S \quad \Rightarrow \quad G \equiv \frac{I(L)}{I(0)} = 1 + g_0 L \frac{I_S}{I(0)} \to 1 \quad \text{for} \quad I(0) \gg I_S. \tag{8.52}$$

Along with increasing input intensity, the amplification approaches one in the saturation range, since the increase in intensity decreases relative to the input signal. In Fig. 8.9, the increase in intensity is represented in the active medium; the three amplification regions are sketched in.

8.4 Laser Output Power and Efficiency

8.4.1 *Available Amplification Power*

On the basis of the expressions worked out in the last section for the signal amplification, the maximum amount of power can be determined from a laser medium of a specific length in a simple passage. The feedback mechanism is not yet factored in here; described, rather, is the behavior of a saturable amplifier.

The output taken from the laser medium, with reference to the cross-section area, is given by the growth in intensity:

$$I_{\text{extr}} = I(L) - I(0) = I_S \cdot \ln\left(\frac{G_0}{G}\right). \tag{8.53}$$

I_{extr} Output extracted from the laser medium per cross-sectional area.

For small input intensities, the amplification G corresponds to the small-signal amplification G_0. Output intensity and extracted power are then approximately the same and much larger compared to the input intensity. At higher input intensities, the amplifier begins to saturate. The extracted power reaches its limit value when the amplifier is completely saturated:

$$I_{\text{extr,max}} = \lim_{G \to 1} I_{\text{extr}} = I_S \ln G_0 = I_S g_0 L. \tag{8.54}$$

$I_{\text{extr,max}}$ Maximally extractable power per cross-sectional area.

This is the maximum power to be taken from the laser medium, with reference to the cross-sectional area. In this borderline case, the amplification returns to approach nearly one, since a very high input signal level is necessary to drive the amplifier completely into saturation. To interpret it better physically, Eq. 8.54 can be converted to

$$I_{\text{extr,max}} = g_0 L \cdot \frac{\hbar\omega c}{B_{21}\tau_{\text{eff}}} = L\frac{D_0\hbar\omega}{\tau_{\text{eff}}} \quad \Rightarrow \quad \frac{I_{\text{extr,max}}}{L} \equiv \frac{P_{\text{extr,max}}}{V} = \frac{D_0\hbar\omega}{\tau_{\text{eff}}}, \tag{8.55}$$

$P_{\text{extr,max}}$ Maximally extractable power from the laser medium

while using the definitions of the small-signal amplification and the saturation intensity. This means the maximum power that can be taken from the laser medium per volume is given by the energy stored in the small-signal or input inversion, $D_0\hbar\omega$, divided by the lifetime of the signal amplification τ_{eff}. The *extraction efficiency* is defined as the relationship of the extracted power to the maximally extractable power:

$$\eta_{\text{extr}} = \frac{I_{\text{extr}}}{I_{\text{extr,max}}} = \frac{\ln G_0 - \ln G}{\ln G_0} = 1 - \frac{\ln G}{\ln G_0}. \tag{8.56}$$

η_{extr} Extraction efficiency.

The main problem when applying nonlinear amplifiers lies in that a higher extraction efficiency can only be reached at low amplification and, vice versa, high amplification only at a low efficiency. Using feedback coupling and converting the amplifier into an oscillator can partially rectify this problem.

8.4.2 *Laser Output Power*

In the laser resonator, the intensity—present at the longitudinal coordinate z—is composed of a contribution of the forth and back travelling waves (Fig. 8.10):

$$I(z) = I_+(z) + I_-(z). \tag{8.57}$$

I_+, I_- Intensity of the forth and back travelling wave

The boundary conditions at the resonator mirrors at $z = 0$ and $z = L$, respectively,

$$I_+(0) = R_1 I_-(0), \quad I_-(L) = R_2 I_+(L) \tag{8.58}$$

R_1, R_2 Degree of reflection at the resonator mirror

have to be fulfilled. For both waves, a differential equation, Eq. 8.45, is applied separately,

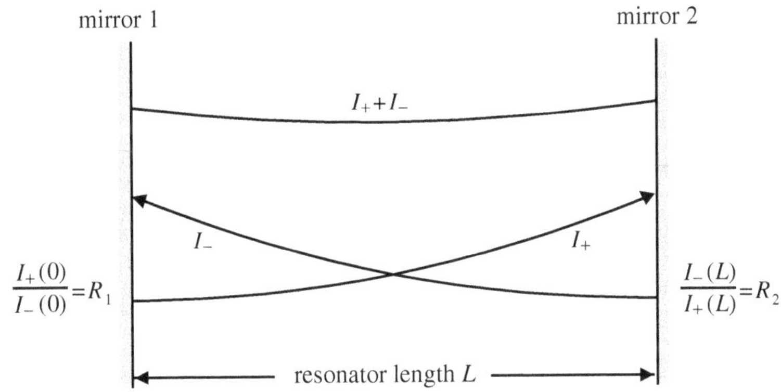

Fig. 8.10 Intensity in the resonator, for the advancing and returning waves

$$\frac{dI_+}{dz} = [g(z) - a]I_+, \quad \frac{dI_-}{dz} = -[g(z) - a]I_-, \tag{8.59}$$

with the nonlinear amplification coefficients

$$g(z) = \frac{g_0}{1 + [I_+(z) + I_-(z)]/I_S} \tag{8.60}$$

and the loss coefficients a. The total intensity causes the saturation of the amplification. The loss coefficient a describes the absorption losses in the medium.

If one assumes the mirror outcouples marginally, hence with reflection coefficients R_1 and R_2 approaching one, it can also be assumed that the intensity is nearly constant over the length of the resonator:

$$I_+(z) \approx I_-(z) \approx I_{\text{zirk}}. \tag{8.61}$$

I_{circ} is the intensity circulating in both directions in the resonator. This ansatz immediately results in the nonlinear amplification coefficient also being approximately independent of the position in the resonator:

$$g \approx \frac{g_0}{1 + 2I_{\text{zirk}}/I_S}. \tag{8.62}$$

In this way, the differential equations, Eq. 8.59, can be solved by an exponential ansatz, and by using the boundary conditions, Eq. 8.58, can be summarized to the condition for stationary operation:

$$\left.\begin{aligned}I_+(L) &= I_+(0)\mathrm{e}^{(g-a)L}\\ I_-(0) &= I_-(L)\mathrm{e}^{(g-a)L}\end{aligned}\right\} \quad\Rightarrow\quad I_-(0) = R_1R_2I_-(0)\mathrm{e}^{2L(g-a)} \quad\Rightarrow \qquad (8.63)$$
$$R_1R_2\mathrm{e}^{2L(g-a)} = 1.$$

If the expression from Eq. 8.62 is now inserted for g, it can be solved according to I_{circ}:

$$I_{\mathrm{circ}} = \left[\frac{2Lg_0}{2La - \ln(R_1R_2)} - 1\right]\cdot\frac{I_S}{2} \quad=\quad (r_{\mathrm{th}} - 1)\cdot\frac{I_S}{2}. \qquad (8.64)$$

The *threshold value ratio* r_{th} indicates the relationship of the unsaturated gain per resonator circulation to the entire circulation losses. Thus, it represents a gauge for how far the laser has been driven beyond the laser threshold by strong pumping; the value $r_{\mathrm{th}} = 1$ corresponds to operation exactly on the threshold. The expression is valid only above the threshold, since for $r_{\mathrm{th}} < 1$ the circulating intensity would become negative.

The useful *outcoupled laser power* is the part of the output circulating in the resonator, which has been outcoupled at the front mirror,[6]

$$I_{\mathrm{aus}} = T_2I_{\mathrm{zirk}} \approx T_2\left[\frac{2Lg_0}{2La + T_1 + T_2} - 1\right]\cdot\frac{I_S}{2}, \qquad (8.65)$$

I_{out} Outcoupled power
$T_{1,2} = (1 - R_{1,2})$ Degree of transmission of the mirror

whereas the approximation

$$\ln(R_1R_2) = \ln[(1 - T_1)(1 - T_2)] \approx \ln(1 - T_1 - T_2) \approx -T_1 - T_2, \qquad (8.66)$$

is used for small degrees of transmission T_1 and T_2. In Fig. 8.11 the course of the outcoupled laser power is plotted over the degree of transmission T_2 for different small-signal amplification coefficients, and in Fig. 8.12 for different absorption coefficients.

[6]The concepts power and intensity are used here synonymously, since—in these calculations—the intensity can be converted into output simply by multiplying it with the cross-sectional area of the laser medium; effects dependent upon transversal coordinates are neglected.

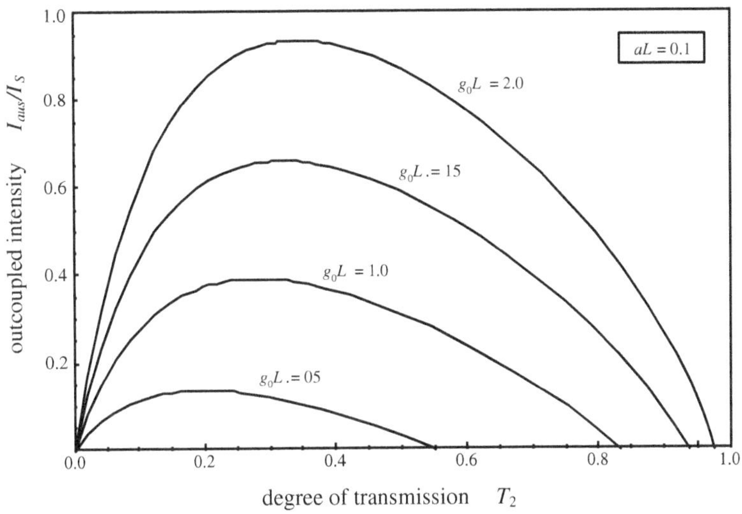

Fig. 8.11 The outcoupled laser intensity, plotted over the transmission degree T_2 of the outcoupling mirror. $T_1 = 0.05$ was assumed for the degree of transmission of the second mirror. The image is not based on the approximation for small degrees of transmission according to Eq. 8.66, but rather on the expression for I_{circ} according to Eq. 8.64

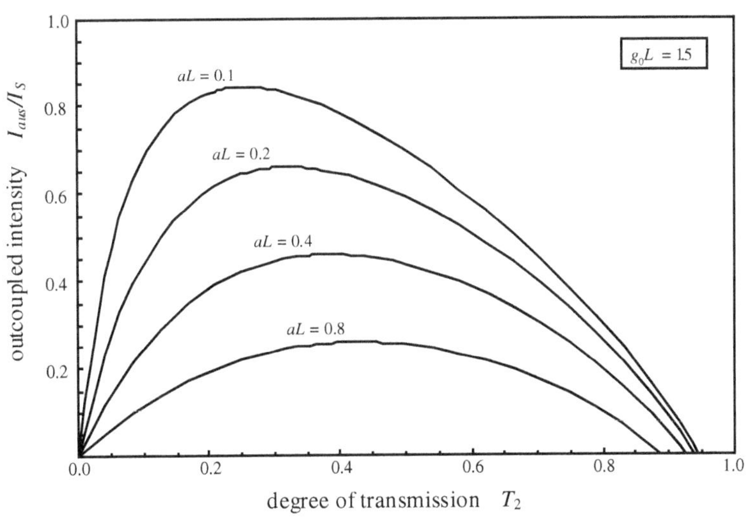

Fig. 8.12 As in Fig. 8.11, but for a constant amplification coefficient and different absorption coefficients

8.4.3 *Optimal Degree of Outcoupling and Optimal Laser Power*

Apparently there is, as seen in Fig. 8.11, a value concerning the outcoupled power for the degree of transmission of the front mirror. The determination of this optimal value occurs according to the common procedures. If one differentiates Eq. 8.65 according to T_2 and finds the roots,

$$T_{2,\mathrm{opt}} = \sqrt{2Lg_0(2La+T_1)} - (2La+T_1) = \left[\sqrt{\frac{2Lg_0}{2La+T_1}} - 1\right] \cdot (2La+T_1) \tag{8.67}$$

$T_{2,\mathrm{opt}}$ Optimal degree of transmission of the front mirror.

results. The maximum laser output power follows, in turn, by inserting the optimal degree of transmission in the output equation, Eq. 8.65. To simplify, the losses are neglected because of the radiation that emits through the rearward resonator mirror,

$$T_1 = 0 \quad \Rightarrow \quad I_{\mathrm{out,max}} = g_0 L \cdot I_S \cdot \left(1 - \sqrt{\frac{a}{g_0}}\right)^2 = I_{\mathrm{extr,max}} \cdot \left(1 - \sqrt{\frac{a}{g_0}}\right)^2, \tag{8.68}$$

$I_{\mathrm{out,max}}$ Maximum, outcoupled intensity

due to Eq. 8.54. The power outcoupled at optimal transmission is, thus, proportional to the maximally extractable power, and the proportionality coefficient depends only on the ratio of the small-signal amplification to the absorption in the medium. The ratio of outcoupled power to maximally extractable power is called the *power-extraction efficiency* η_{pow}:

$$\eta_{\mathrm{pow}} = \frac{I_{\mathrm{out}}}{I_{\mathrm{extr,max}}} \quad \Rightarrow \quad \eta_{\mathrm{pow,max}} = \frac{I_{\mathrm{out,max}}}{I_{\mathrm{extr,max}}} = \left(1 - \sqrt{\frac{a}{g_0}}\right)^2. \tag{8.69}$$

The maximum power-extraction efficiency is a measure for how well the power can be utilized from the laser medium. It approaches one when the small-signal amplification is much larger than the absorption losses in the medium.

8.4.4 *Laser Efficiency*

For industrial applications, the *total efficiency* of the laser system is, above all, significant:

$$\eta_{\text{total}} = \frac{P_{\text{out}}}{P_{\text{in}}}. \tag{8.70}$$

In this context, P_{out} identifies the total power of the outcoupled laser beam and P_{in} the total power fed into the laser system.

The input power is normally made available in electrical form; it is converted into the external pump rate P^* in a manner specific for each pump process, which with atoms from the basic level E_0 are excited into the upper pump level E_3 (cf. Fig. 8.6). Here the first losses appear, which are described by the *external pump efficiency* $\eta_{P,\text{ext}}$:

$$\eta_{P,\text{ext}} = \frac{P^* \cdot \Delta E_{\text{Pump}}}{P_{\text{in}}} = \frac{P^*(E_3 - E_0)}{P_{\text{in}}}. \tag{8.71}$$

The pump process and, accordingly, the external pump efficiency can be composed of several individual steps.

From the upper pump level, the atoms relax into the upper laser level E_2. Several of the atoms pass over into lower energy states immediately such that they are lost for the laser process. This is incorporated by using the *internal pump efficiency* η_P (cf. Eq. 8.7),

$$\eta_P = \frac{P}{P^*}, \tag{8.72}$$

whereas the internal pump rate P indicates how many atoms per unit of time make it into the upper laser level from the pump level. From this position, the laser transition takes place. The energy difference between the laser levels E_1 and E_2 is, however, smaller than the energy difference $E_3 - E_0$ the pump process has to provide so that, in turn, an effective energy loss is recorded for every radiating atom. This is indicated by the quantum efficiency η_Q:

$$\eta_Q = \frac{\Delta E_{\text{Laser}}}{\Delta E_{\text{Pump}}} = \frac{\hbar\omega}{E_3 - E_0}. \tag{8.73}$$

ω Angular frequency of laser emission

Now, however, atoms from the upper laser level also pass into lower energy states via other processes than stimulated emission, for example by spontaneous emission. These losses are commonly formulated as the *photon yield* $\eta_{\hbar\omega}$

$$\eta_{\hbar\omega} = \left(\frac{\mathrm{d}N_2}{\mathrm{d}t}\right)_{\text{stim}} \Big/ \left(\frac{\mathrm{d}N_2}{\mathrm{d}t}\right)_{\text{total}} = \frac{B_{21}\rho}{A_{21} + A_{2*} + B_{21}\rho}. \tag{8.74}$$

In this manner the photon yield is, however, difficult to determine since the photon density ρ also enters the equation. On the other hand, it has already been determined, as seen in Sect. 8.4.1, which part of the power fed by pumping can be best used to amplify the radiation: the maximum amplifier output available was indicated by $I_{\text{extr,max}}$ (cf. Eq. 8.54). Therefore, the *maximum attainable photon yield* can also be notated as

$$\eta_{\hbar\omega} = \frac{P_{\text{extr,max}}}{P \cdot \hbar\omega} = \frac{I_{\text{extr,max}} \cdot A_{\text{cross}}}{P \cdot \hbar\omega} \tag{8.75}$$

A_{cross} Cross-sectional area of the laser medium.

All further losses arising from the amplification and feedback in the resonator have already been summarized in the power-extraction efficiency η_{pow} (Eq. 8.69):

$$\eta_{\text{pow}} = \frac{I_{\text{aus}}}{I_{\text{extr,max}}} = \frac{P_{\text{aus}}}{P_{\text{extr,max}}}. \tag{8.76}$$

The total efficiency of the laser system now results from the product of the individual efficiencies:

$$\eta_{P,\text{ext}}\eta_P\eta_Q\eta_{\hbar\omega}\eta_{\text{pow}} = \frac{P_{\text{out}}}{P_{\text{in}}} \equiv \eta_{\text{total}}. \tag{8.77}$$

If the maximum extractable power is traced back to the rate coefficients by means of Eqs. 8.55 and 8.41, the following holds for the photon yield using Eq. 8.75:

$$P_{\text{extr,max}} = \frac{A_{10} - A_{21}}{A_{10} + A_{2*}} \hbar\omega P \quad \Rightarrow \quad \eta_{\hbar\omega} = \frac{A_{10} - A_{21}}{A_{10} + A_{2*}}. \tag{8.78}$$

Normally the transition from the lower laser level in the ground state occurs very quickly as compared to the spontaneous depopulation of the upper laser level. Then $A_{10} \gg A_{21}, A_{2*}$, and the photon yield is almost one.

8.5 Hole Burning and Multimode Operation

Until now spectral properties of the amplification have not been discussed. The position of the laser line has been given in the form of the laser transition frequency ω, and line widths and frequency dependencies have been neglected. To understand the laser process and its principle properties, this model will suffice. Due to the limitations previously mentioned, however, it is only valid for single-mode operation in the strictest sense. Therefore, this section shall mainly serve to represent the causes of multimode operation.

8.5.1 *Ideally Homogeneously Enhanced Laser Line*

In the previous sections it was assumed that the laser energy level is at any level of sharpness and the radiation of the respective transitions completely monochromatic. In fact the transition lines always, however, exhibit a finite line width, which can materialize in different ways.

Every transition exhibits the *natural line* width. This width is the consequence of the energy uncertainty of the level due to its finite life time. The line has, thus, a Lorentz profile (cf. Sect. 7.4.5):

$$f_n(\omega) = \frac{\Delta\omega_n}{(\omega - \omega_{21})^2 + \Delta\omega_n^2}, \quad \Delta\omega_n = \frac{1}{\tau_2} + \frac{1}{\tau_1}. \tag{8.79}$$

$\tau_{1,2}$ Life time of the states
$\Delta\omega_n$ Natural line width
ω_{21} Central transmission frequency.

Since this kind of line broadening is independent of inhomogeneities in the medium, it is called the *homogenous line broadening.*

The spectral line form passes directly into the nonlinear amplification coefficient g (Eq. 8.60), which means

$$g \equiv g(\omega) \sim f_n(\omega) \tag{8.80}$$

in the case of the homogenous line broadening. When simply passing through an active medium of the length L, the amplification G is defined as

$$G(\omega) = \frac{I(L)}{I(0)} = \exp\{g(\omega) \cdot L\}, \tag{8.81}$$

such that the line shape function $f_n(\omega)$ now appears in the exponent. This means that the amplification decreases much more quickly with increasing imbalance $\omega - \omega_{21}$ than the atomic line itself. The amplification bandwidth of the medium is, therefore, always smaller than the atomic line width $\Delta\omega_N$, and continues to fall as amplification grows. This effect is also called *gain narrowing.*

As a rule the amplification curve of the active medium is much wider than the line width of the laser resonator (cf. Sect. 6.2.4); mostly the amplification bandwidth is even significantly larger than the difference frequency of neighboring, longitudinal resonator modes. This means that many modes lie within the amplification bandwidth and can, in principle, be amplified. In an ideal laser medium with homogenous line broadening, the line form is constant and the same for all atoms of the medium. If the amplification is increased by stronger pumping, the entire spectral amplification curve is shifted upwards, until those modes that lie closest to the mean transmission frequency ω_{21} reach the laser threshold (Fig. 8.13).

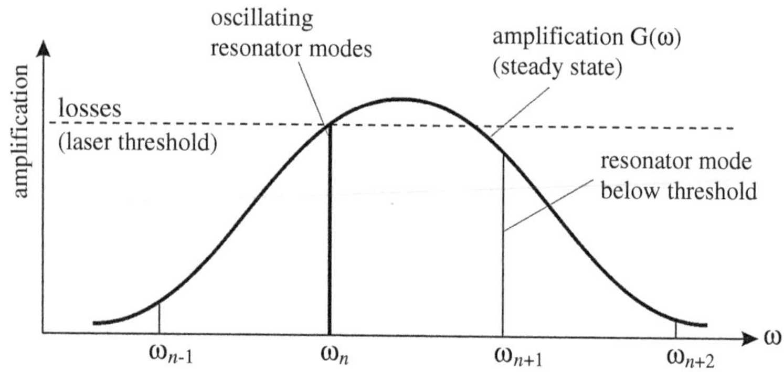

Fig. 8.13 In an ideal laser medium with homogenous line broadening, only one mode can reach the laser threshold

For these modes, the amplification and the losses balance each other out; these modes can oscillate, whereas the losses still dominate and lead to damping for all the other modes.

Even when the pump rate is increased further, the amplification profile cannot be raised further so that the next resonator mode begins to oscillate. This is not possible in a stationary operating state because the amplification would dominate the losses for the first mode and, thus, its amplitude would continue to increase: if a mode has reached the laser threshold, the population inversion remains constant on its saturation value (cf. Sect. 8.3.1) and the amplification cannot continue to increase (cf. Sect. 8.3.3). The additional pump power fed in is directly converted into optical output power. In an ideal laser with homogenous line broadening, only a single mode can begin to oscillate in stationary operation. In real lasers, however, the validity of this conclusion is limited by practical aspects. A significant mechanism that can lead to multimode oscillations in lasers with homogenous line broadening is *spatial hole burning*, described in Sect. 8.5.4. Nonetheless, practice has also established that lasers with homogenous broadening more likely tend toward monomode operation. Semiconductor lasers and most of the solid-state lasers are examples of this.

8.5.2 Homogeneous Line Broadening

The natural line width represents the minimum broadening of atomic emission lines. It can only be observed in motionless atoms isolated from each other, thus, for example, in cold, thin gases. If more atoms are brought together more densely, the interaction of the atoms with each other leads to significant line broadening. An important contribution to this is the *collision or pressure broadening*. Collisions between atoms contribute to the line broadening via two mechanisms:

- *Inelastic collisions* with excited atoms can lead to non-radiating transitions, which means the excited atom passes over into a lower energy state due to the collision, but without emitting radiation. Thereby, the effective lifetime of the excited level is shortened and thus the line width increased.
- *Elastic collisions* during the emission process can lead to phase jumps in the emitted pulse of waves. These phase jumps also result in a line broadening, although the lifetime is not shortened.

Due to the collision broadening the line width increases approximately linearly along with the partial density or rather the gas pressure.

While the former formulation of the collision broadening primarily applies to gas lasers, there are also equivalent processes in solid-state laser media. In solid bodies, the atoms are bound to fixed positions. Therefore, they do not collide directly as in gases, but rather make different oscillation movements around their resting positions. The interaction with these oscillations then leads to the broadening mechanism described above.[7]

8.5.3 Inhomogeneous Broadening and Spectral Hole Burning

In addition to the homogeneous line broadening named in the last section, there are different mechanisms that lead to an *inhomogeneous broadening* of the spectral transition line. With inhomogeneous broadening, external influences lead to a scattering of transition frequencies ω_{21} of the atoms. In this case the broadening is of a static nature; the line form of inhomogeneously broadening lines has, therefore, a Gaussian profile. Examples for inhomogeneous line broadening are the *Doppler broadening* of gas lasers and the *line broadening in amorphous solid bodies*:

- A cause for the Doppler broadening is the Doppler effect: the emission line of an atom moved relative to the observer is shifted dependent upon the movement

[7]In the quantum theory of the solid body, such lattice vibrations are called *phonons* and are treated similar to photons as particles. The interaction of the atoms with the phonons is then seen as a collision between atom and phonon.

direction and velocity compared to the frequency at the state of rest. Above all, this effect appears in gases as a consequence of the thermal atom movement. The Doppler line width connected to the thermal movement amounts to

$$\Delta\omega_d = \frac{2\omega_{21}}{c}\sqrt{\frac{2\ln 2\, k_B T}{m}} \tag{8.82}$$

$\Delta\omega_d$ Doppler line width
T Gas temperature
m Mass of the gas particles

- Amorphous solid bodies (*glass*) do not exhibit a regular crystal structure, but rather are comparable to a frozen liquid in their structure: atoms and molecules lie in different arrangement and orientation to each other. Each laser-active atom embedded in this mass comes across, therefore, slightly different surroundings, so that the atoms exhibit slightly deviating energy levels and, thus, emit at correspondingly different frequencies. Since the entire emission line of the solid body is made up of individual emission lines, a line broadening results.
This kind of broadening is clearly visible in a comparison between Nd:YAG and Nd:Glass lasers. In the first case, the laser-active neodymium atoms are embedded in a YAG crystal (Yttrium Aluminum Garnet), and the emission line is broadened homogeneously. The typical line width amounts to approx. $\Delta\nu = 1.2 \times 10^{11}$ Hz. In the second case, a significantly stronger, inhomogeneous line broadening appears with approx. $\Delta\nu = 7.5 \times 10^{12}$. From the different line broadening it also follows that the maximum emission in the center of the line for Nd:YAG is much larger than for Nd:Glass at the same Nd concentration.

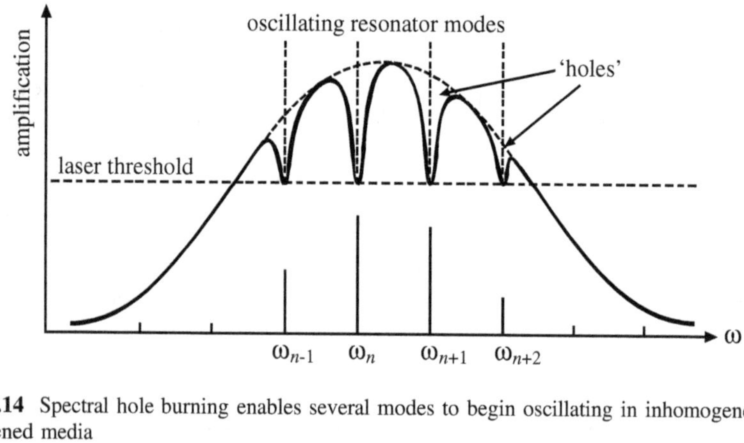

Fig. 8.14 Spectral hole burning enables several modes to begin oscillating in inhomogeneously broadened media

For inhomogeneous broadening, the amplification is carried in every spectral region of the line by only a part of the atoms distributed statically in the volumes. Every axial mode of the resonator can, therefore, when the amplification exceeds the losses, bring about—independent of the other modes—the saturation in the part of the atoms that are in resonance with the mode frequency (Fig. 8.14). Therefore, every mode can cause a "hole" in the spectral amplification curve. This effect is thus called *spectral hole burning*. As a consequence of spectral hole burning, the oscillation build-up of several laser modes is possible in laser media with inhomogeneous line broadening.

8.5.4 Spatial Hole Burning

Independent of the kind of line broadening, a spatial inhomogeneity of the amplification can lead to multimode operation. In a common laser resonator, the axial resonator modes correspond to standing waves: exactly the number of half wavelengths λ_n given by the mode order fit in the resonator length L (cf. Sect. 6.1.1):

$$n\frac{\lambda_n}{2} = L.$$

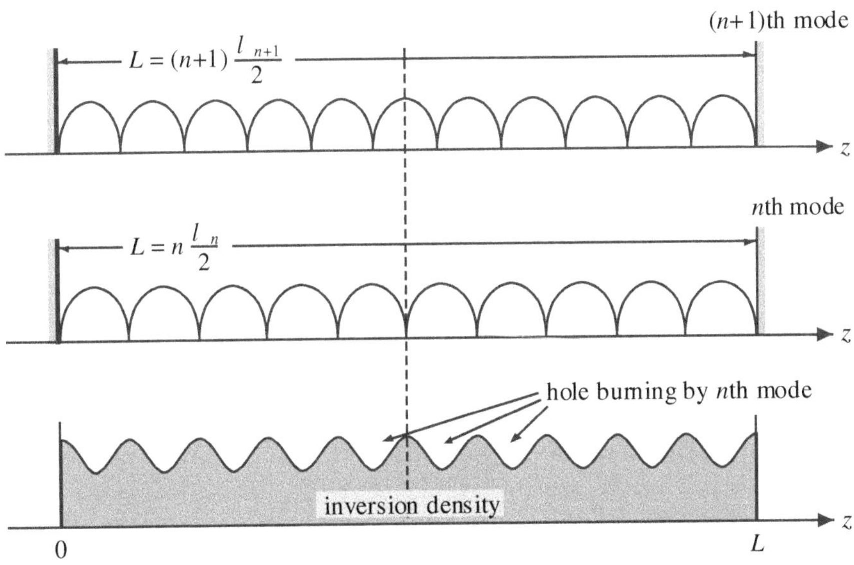

Fig. 8.15 Spatial hole burning. Longitudinal modes of different orders saturate different regions of the laser medium

Standing waves are characterized, however, by fixed nodes and antinodes of the field. This means there are fixed regions with lower mean intensity in which these modes do not reach saturation. At these positions, the mode with the next higher or lower order can, however, exhibit a wave antinode (Fig. 8.15). In this way, the amplification of neighboring modes can take place in different regions of the medium in terms of emphasis. The decoupling of the amplification for the individual modes enables several modes to begin oscillating simultaneously.

8.6 Nonstationary Behavior and Pulse Generation

8.6.1 Spiking

Until now only static solutions have been examined, which means the pump rate may only change in periods of time which are large compared to the time constants of the system. The main time scale in this case is the lifetime of the upper laser level. When the pump rate oscillates quickly, the stationary equilibrium solution is no longer valid. This is especially valid when the system is turned on, when the pump rate bounces up from zero to a fixed value. Typically, population inversion and photon density show a transient behavior, as represented in Fig. 8.16. Due to

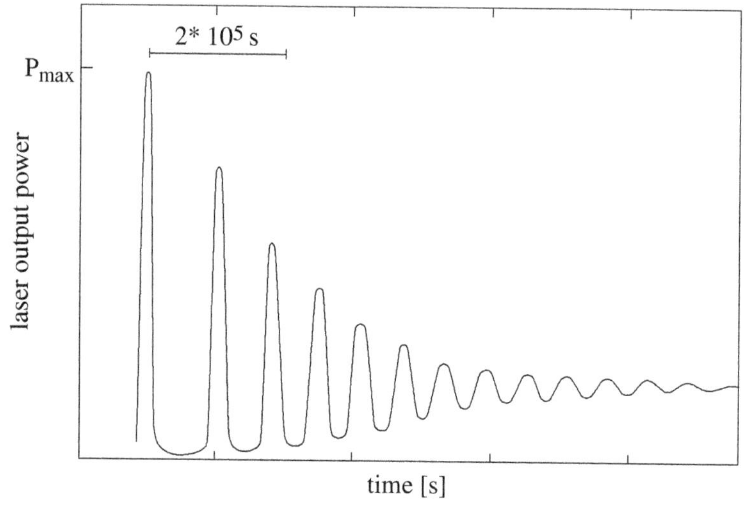

Fig. 8.16 Measured transient behavior (*spiking*) of the output power of a stable ground mode laser. The figure represents the average of over 20 individual measurements

the "spikes" appearing at this time in the output power of the laser, this behavior is called *spiking*.[8]

The starting point for the description of spiking are the rate equations (Eq. 8.19) for high pump rates and with self-excitation, which means with the simplifications

$$K' = 0, \quad FD' \approx 0, \quad \delta \approx 0. \tag{8.83}$$

For this reason, the rate equations are

$$\begin{aligned} \frac{\mathrm{d}}{\mathrm{d}\tau} D' &= \alpha P' - (1+\rho')D' \\ \frac{\mathrm{d}}{\mathrm{d}\tau} \rho' &= D'\rho' - \alpha\rho'. \end{aligned} \tag{8.84}$$

Solutions for the stationary case are

$$\frac{\mathrm{d}}{\mathrm{d}\tau} D' = \frac{\mathrm{d}}{\mathrm{d}\tau} \rho' = 0 \quad \Rightarrow \quad D_0' = \alpha, \rho_0' = P' - 1. \tag{8.85}$$

This agrees with the solutions for high pump rates from Sect. 8.3.1 if only the term proportional to F in Eq. 8.33 is neglected, which, nonetheless, remains one. Back substitution leads to:

$$D_0 = \frac{\beta}{B_{21}}, \quad \rho_0 = \frac{P}{\beta} - \frac{A_{21}}{B_{21}}. \tag{8.86}$$

For time-dependent *variations* around the stationary equilibrium solutions, the following ansatz according to Dunsmuir can be used:

$$\rho(t) = \rho_0(1+\varepsilon(t)), \quad D(t) = D_0(1+\eta(t)) \quad \text{mit} \quad |\varepsilon(t)|, |\eta(t)| \ll 1. \tag{8.87}$$

Inserting it into the rate equations, Eq. 8.84, and neglecting the products from ε and η lead to

$$\begin{aligned} \dot{\eta} &= -\frac{B_{21}}{\beta} P\varepsilon + A_{21}\varepsilon - \frac{B_{21}}{\beta} P\eta \\ \dot{\varepsilon} &= \beta\eta \end{aligned} \tag{8.88}$$

By renewed derivation according to the time, ε can be eliminated from the first equation, and the equation becomes the typical damped oscillation equation:

$$\begin{aligned} \ddot{\eta} + a\dot{\eta} + b\eta &= 0, \\ \dot{\varepsilon} &= \beta\eta \end{aligned} \quad \text{with} \quad a = \frac{B_{21}P}{\beta}, b = B_{21}P - A_{21}\beta \tag{8.89}$$

8.

This is solved using a complex exponential ansatz:

$$\begin{aligned} \tilde{\eta} = \tilde{\eta}_0 e^{i\omega t - \gamma t} \quad &\Rightarrow \quad (i\omega - \gamma)^2 + a(i\omega - \gamma) + b = 0 \\ &\Leftrightarrow \quad -\omega^2 + \gamma^2 - 2i\omega\gamma + ia\omega - a\gamma + b = 0. \end{aligned} \tag{8.90}$$

Since real and imaginary parts each have to become zero individually, Eq. 8.90 can be separated into two equations for ω and γ:

$$\begin{aligned} -2i\omega\gamma + ia\omega = 0 \quad &\Rightarrow \quad \gamma = \frac{a}{2} \\ -\omega^2 + \gamma^2 - a\gamma + b \quad &\Rightarrow \quad \omega^2 = b - \frac{a^2}{4}. \end{aligned} \tag{8.91}$$

Only the real part of the solution is physically relevant:

$$\eta(t) = \Re(\tilde{\eta}) = \eta_0 \cos \omega t \, e^{-\gamma t}. \tag{8.92}$$

It describes a dampened oscillation at the frequency ω. ε is connected to η by a simple time derivation (Eq. 8.88), and therefore a phase shifted, harmonic oscillation results for ε:

$$\begin{aligned} &\dot{\varepsilon} = \beta\eta \quad \Rightarrow \quad \tilde{\varepsilon} = \tilde{\varepsilon}_0 e^{i\omega t - \gamma t}, \\ &\text{with} \quad \tilde{\varepsilon}_0 = \frac{\beta}{i\omega - \gamma}\tilde{\eta}_0 = \frac{\beta}{\sqrt{\omega^2 + \gamma^2}} e^{-i\phi}\tilde{\eta}_0, \quad \phi = \arctan\frac{\omega}{\gamma}. \end{aligned} \tag{8.93}$$

For the real ε, from this it follows

$$\varepsilon(t) = \varepsilon_0 \cos(\omega t - \phi) e^{-\gamma t}, \quad \varepsilon_0 = \frac{\beta\eta_0}{\sqrt{\omega^2 + \gamma^2}}, \tag{8.94}$$

which is a harmonic and equally dampened oscillation, with a phase shift of φ. The phase shift depends upon the relationship of the frequency ω to the damping coefficient γ: for weak damping, with $\gamma \to 0$, the phase difference approaches $\pi/2$.

In total, the following transient oscillations result for the population inversion and the photon density:

$$\begin{aligned} D(t) &= D_0(1 + \eta_0 \cos \omega t e^{-\gamma t}) \\ \rho(t) &= \rho_0(1 + \varepsilon_0 \cos(\omega t - \phi) e^{-\gamma t}) \end{aligned} \quad \text{with} \quad \begin{aligned} \omega &= \sqrt{B_{21}P - A_{21}\beta - \gamma^2}, \\ \gamma &= \tfrac{B_{21}P}{\beta}, \ \phi = \arctan\tfrac{\omega}{\gamma}. \end{aligned} \tag{8.95}$$

In Fig. 8.17, the numerical calculation of the transient behavior of a ruby laser is represented. As seen previously, the ansatz according to Dunsmuir was used to make these calculations.

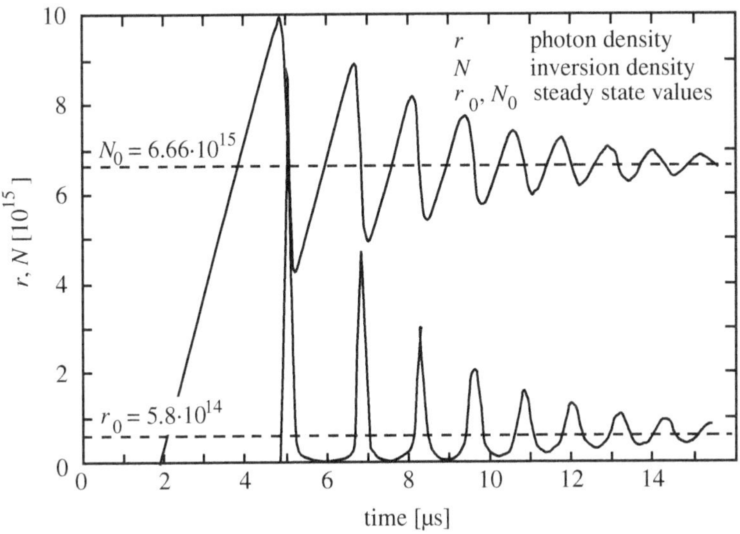

Fig. 8.17 Numerically calculated transient behavior of a ruby laser according to Dunsmuir

Great significance is placed upon spiking in practice: every quick change in the operation of the laser system leads to variations of the system parameters and can, therefore, negatively influence the desired processing result. What needs to be considered, above all, is that the transient behavior of each laser mode is different. The linear superpositioning of the modes in the outcoupled signal leads, thus, to a superposition of numerous, nonlinear disturbances and to a chaotic emission behavior in the end. Such behavior is inappropriate for applications requiring high precision.

By contrast, the spiking shown here can be seen only in single-mode lasers in its pure form. There the spiking can be cultivated and utilized by canceling the damping γ of the oscillations through modulating the pump rate or the resonator losses appropriately and periodically: the results are regular pulses in the distance $T = 2\pi/\omega$. The distance and the amplitude of the pulses can be calibrated using the pump rate and the resonator losses. These thoughts lead to the basic idea of pulse producing using the Q-switch process.

8.6.2 Nonstationary Pulse Generation: The Q-Switch Laser

Q-switching is a procedure to generate regular, short laser pulses with high intensity. Pulse lengths of several nanoseconds and pulse power of up to several gigawatts can be reached using this procedure. The principle of q-switching was proposed shortly after the first laser was realized; today it counts among the standard procedures of laser technology.

When generating pulses, q-switching takes advantage of the fact that in the laser medium, energy is stored and can be retrieved with delay. The functional principle is already expressed by the relationship: the quality of the optical resonator is switched between low and high values. The quality is defined as the ratio of the energy stored in the resonator to the dissipated power:

$$Q = \Omega \frac{E}{\dot{E}}. \tag{8.96}$$

$\Omega = 2\pi c/2L$ Resonator circulation angular frequency
E Energy stored in the resonator
$\dot{E}$ Temporal energy change (dissipated power)

The quality can be controlled via the resonator losses (Fig. 8.18). A resonator of high quality is characterized by low losses. Since, according to Eq. 8.36, the threshold to use laser activity is directly proportional to the loss factor β of the resonator, the threshold for a resonator of high quality is particularly low.

With the q-switch, the quality of the resonator is first strongly diminished; due to this, the laser threshold rises so much that laser activity is no longer reached. This leads, however, to the upper laser level emptying only slowly through spontaneous emission; with undiminished strong pumping, the population inversion becomes, therefore, very large. When the inversion has reached its maximum, the quality of the resonator is scaled up again. The laser threshold sinks abruptly to a very low value and the total population inversion collected in the build-up phase is decomposed over the laser transition in a short period of time. A short, very intensive laser pulse (giant pulse) is produced in the process. Subsequently the process can begin anew. This way, regular, successive, pulses rich in energy are produced. The temporal process of a cycle is represented in Fig. 8.19.

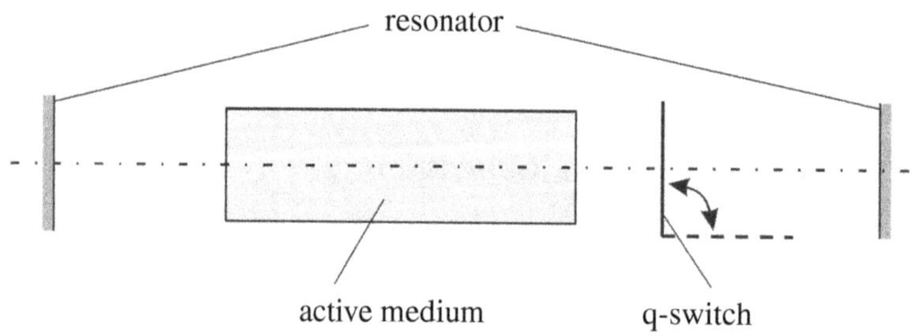

Fig. 8.18 Schematic construction of a q-switched laser. The q-switch interrupts the feedback through the resonator, i.e., the resonator is deactivated and the losses rise to a maximum

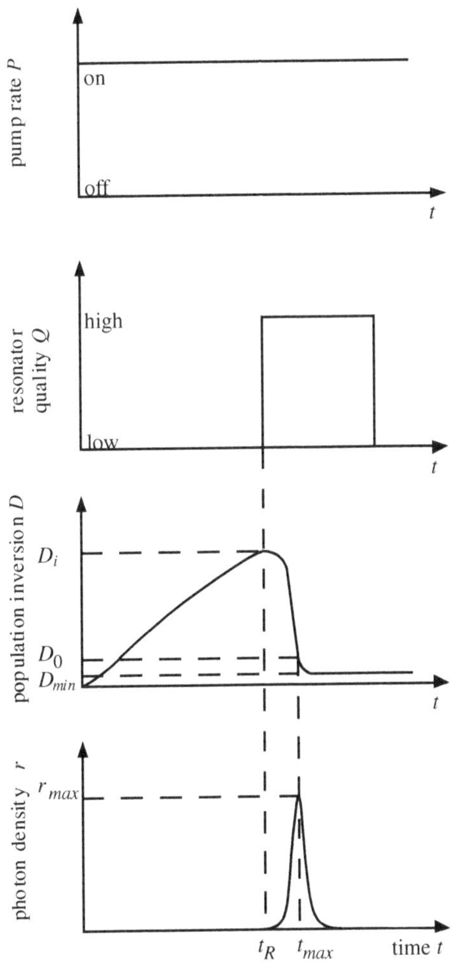

Fig. 8.19 An explanation of the temporal process in a laser using the q-switch procedure

The same rate equations are used as those for spiking (cf. Eq. 8.84) to describe q-switch operation. When Eqs. 8.17 and 8.18 are taken into consideration for the unnormalized values, the rate equations read as follows:

$$\begin{aligned} \dot{D} &= P - B_{21}D\rho - A_{21}D \\ \dot{\rho} &= B_{21}D\rho - \beta\rho. \end{aligned} \tag{8.97}$$

Now, approximate solutions are determined for the population inversion and the photon density in the different phases of q-switching. The point of time at which the resonator is activated, which means when the quality switch is opened, is designated $t = t_R$. The photon density, and thus the laser pulse, reaches its maximum at $t = t_{max}$ (cf. Fig. 8.19).

8.6.2.1 Inversion Build-up ($t < t_R$)

As long as the quality of the resonator is low, laser operation does not take place and the photon density is very low:

$$\rho \approx 0. \tag{8.98}$$

For this reason, the equation for the population inversion can be simplified and solved,

$$\dot{D} \approx P - A_{21}D \quad \Rightarrow \quad D(t) = \frac{P}{A_{21}}\left(1 - \mathrm{e}^{-A_{21}t}\right) = P\tau_{21}\left(1 - \exp\left(-\frac{t}{\tau_{21}}\right)\right), \tag{8.99}$$

with the lifetime of the upper laser level, $\tau_{21} = 1/A_{21}$. The population inversion climbs, until the *maximal inversion* $\tau_{21}P$ is reached asymptotically. Except for the pump rate, the maximum inversion is directly proportional to the lifetime of the upper laser level. A long life time τ_{21} is, therefore, a prerequisite for effective q-switching.

In this context, τ_{21} is called the *storage time* of the medium: it indicates how long the population inversion can be stored in the medium without being decomposed by spontaneous emission. For most of the gas lasers operating in the spectral range, as well as the dye lasers, the storage times amount to only a few nanoseconds. Solid-state lasers are generally more appropriate for quality modulation.

The maximum inversion is only reached asymptotically. In practice, $t_R = 3\tau_{21}$ $-5\tau_{21}$ is selected as the duration of the build-up phase; then, more than 95 % of the maximum inversion has been reached.

8.6.2.2 Start-up Phase ($t > t_R$)

At the time $t = t_R$, the q-switch is opened and the resonator is switched on again. The photon density becomes large very quickly due to the incipient laser process. Therefore, the approximations

$$P \ll B_{21}D\rho, \quad A_{21}D \ll B_{21}D\rho \tag{8.100}$$

are inserted into the rate equations:

$$\begin{aligned}\dot{D} &= -B_{21}D\rho \\ \dot{\rho} &= B_{21}D\rho - \beta\rho.\end{aligned} \tag{8.101}$$

In a short time interval after switching on the resonator, the population inversion changes very little. There, the following holds:

$$D(t) \approx D(t = t_R) \equiv D_i, \quad \rho(t = t_R) \equiv \rho_i. \tag{8.102}$$

ρ_i indicates the photon density that is present due to spontaneous emission before the laser process initiates, in other words, the *noise* of the system. By means of these assumptions, an exponential increase in photon density follows from Eq. 8.101:

$$\rho(t) = \rho_i e^{(B_{21}D_i - \beta)t}. \tag{8.103}$$

From this and with the first equation from Eq. 8.101, the inversion density results in

$$D(t) \approx D_i\left\{1 + \frac{B_{21}}{B_{21}D_i - \beta}\rho_i\left(1 - e^{(B_{21}D_i - \beta)t}\right)\right\}. \tag{8.104}$$

If the photon density continues to increase, the inversion is increasingly decomposed and its change has to be factored into the equation for the photon density. This can be reached in the first approximation by notating the photon density in dependence upon the previously determined expression for the inversion density. For this the approximated rate equations, Eq. 8.101, is converted as follows:

$$\begin{gathered}\left(\frac{dD}{dt}\right)^{-1} = \frac{dt}{dD} = -\frac{1}{B_{21}D\rho} \quad \Rightarrow \quad \frac{d\rho}{dD} = \frac{d\rho}{dt}\cdot\frac{dt}{dD} = \frac{\beta}{B_{21}D} - 1 \equiv \frac{D_0}{D} - 1 \\ \text{with } D_0 = \frac{\beta}{B_{21}}.\end{gathered} \tag{8.105}$$

D_0 is the saturation inversion in laser operation (cf. Eq. 8.35). From the solution of this differential equation, the photon density follows at time t as a function of the population inversion after the resonator is switched on:

$$\rho(t) = D_i - D(t) - D_0 \ln\left(\frac{D_i}{D(t)}\right). \tag{8.106}$$

8.6.2.3 Maximum of the Pulses ($t = t_{max}$)

When the photon density at t_{max} has reached its maximum, then the following holds:

$$\rho(t = t_{max}) = \rho_{max} \quad \Rightarrow \quad \left.\frac{d\rho}{dt}\right|_{t=t_{max}} = 0. \tag{8.107}$$

From Eq. 8.97, thus follows

$$B_{21}D(t_{max})\rho_{max} - \beta\rho_{max} = 0 \quad \Rightarrow \quad D(t_{max}) = \frac{\beta}{B_{21}} = D_0, \tag{8.108}$$

and inserted in Eq. 8.106, the maximum photon density results:

$$\rho_{max} = D_i\left\{1 - \frac{D_0}{D_i} - \frac{D_0}{D_i}\ln\left(\frac{D_i}{D_0}\right)\right\}. \tag{8.109}$$

The quotient D_i/D_0 is called the *threshold value reinforcement*. It indicates how much the inversion lies over the equilibrium inversion in continuous laser operation when the laser pulse is initiated. For large threshold value reinforcements, the factor in parentheses approaches one. Then,

$$\rho_{max} \approx D_i, \tag{8.110}$$

which means nearly the entire inversion is directly converted into laser radiation and a very intensive, short laser pulse is generated.

8.6.2.4 Pulse End ($t > t_{max}$)

Since the inversion is nearly completely converted into laser radiation, it falls quickly under the equilibrium value D_0 for $t > t_{max}$, and since it is still very small, the photon density also collapses (Fig. 8.20):

$$t > t_{max}: \; D = D_{min} < D_0, \quad \rho \to 0. \tag{8.111}$$

When $\rho = 0$ is inserted into Eq. 8.106, the relationship for D_{min} follows:

$$D_i - D_{min} - D_0 \ln\left(\frac{D_i}{D_{min}}\right) = 0 \quad \Leftrightarrow \quad \frac{D_i}{D_0} - \ln\left(\frac{D_i}{D_0}\right) = \frac{D_{min}}{D_0} - \ln\left(\frac{D_{min}}{D_0}\right). \tag{8.112}$$

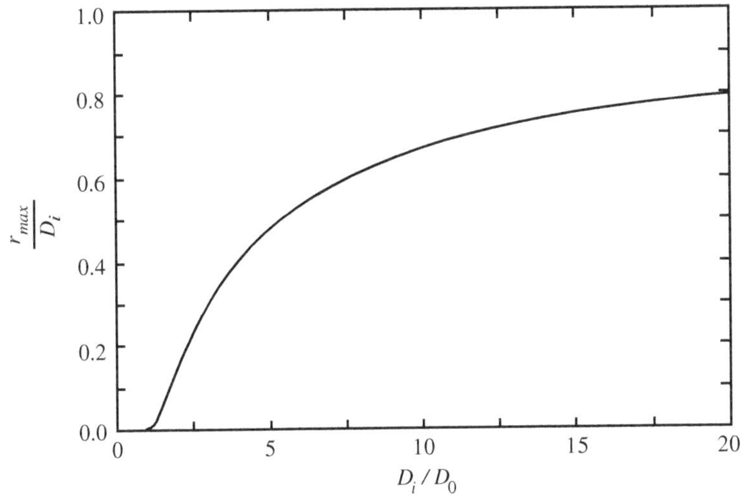

Fig. 8.20 The ratio of maximum photon density to the initial inversion density, plotted over the threshold value reinforcement

This transcendental equation for D_{min} has two solutions: The first solution is $D_{\text{min}} = D_i$ and is eliminated, since it is physically reasonable. From the second solution it follows that

$$D_{\text{min}} < D_0 \quad \text{for} \quad D_i > D_0, \tag{8.113}$$

which the statement in Eq. 8.111 confirms.

From the rate equations, Eq. 8.97, one can derive the decrease in photon density —which means the decreasing flank of the laser pulse—by inserting $D = D_{\text{min}} = \text{const.}$:

$$\dot{\rho} \approx B_{21} D_{\text{min}} \rho - \beta \rho \quad \Rightarrow \quad \rho(t) = \rho_{\text{max}} \mathrm{e}^{-(\beta - B_{21} D_{\text{min}})t}, \quad D_{\text{min}} < D_0 = \frac{\beta}{B_{21}}. \tag{8.114}$$

From the rising edge (Eq. 8.106) and the falling edge (Eq. 8.114) of the laser pulse, its full width at half maximum can be determined while developing the equations by the maximum pulse ρ_{max} and calculating the pulse width for $\rho_{\text{max}}/2$. The approximated half-intensity width results from this:

$$\Delta t_{1/2} \approx \frac{5}{2\beta} \sqrt{\frac{D_0}{\rho_{\text{max}}}} = \frac{5}{2\beta} \sqrt{\frac{D_0}{D_i}}, \tag{8.115}$$

which means the pulse becomes narrower as the resonator quality $Q \sim 1/\beta$ and threshold value reinforcement increase.

The extraction efficiency η_e indicates which part of the energy stored in the initial inversion is converted into laser radiation:

$$\eta_e = \frac{E_{\text{Pulse}}}{E_i}. \tag{8.116}$$

E_{Pulse} Radiation energy emitted in the laser pulse
E_i Energy stored in the initial inversion.

The total energy released in the laser pulse results from the difference of the population inversion before and after the pulse in produced:

$$E_{\text{total}} = (D_i - D_{\min})h\nu. \tag{8.117}$$

Due to internal losses in the medium and resonator, this energy is not completely converted into the pulse energy E_{Pulse}. In praxis these losses can, however, be held to a minimum such that $E_{\text{Pulse}} \approx E_{\text{Total}}$ remains valid. The energy stored from the inversion at the beginning of the q-switch pulse is

$$E_i = D_i h\nu \tag{8.118}$$

such that the extraction efficiency results in

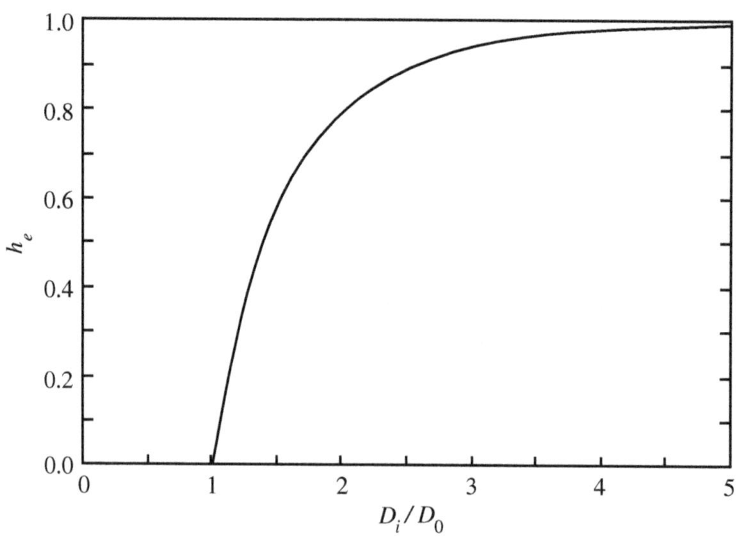

Fig. 8.21 The extraction efficiency as a function of the threshold value reinforcement

$$\eta_e = 1 - \frac{D_{\min}}{D_i} \tag{8.119}$$

Inserting into Eq. 8.112 shows that the extraction efficiency depends solely upon the threshold value reinforcement:

$$\frac{1}{\eta_e} \ln\left(\frac{1}{1-\eta_e}\right) = \frac{D_i}{D_0}. \tag{8.120}$$

The extraction efficiency is already nearly one for small threshold value reinforcements (Fig. 8.21). In general, the threshold value reinforcement is very large; therefore, the efficiency of the q-switch procedure is very high. Nearly the entire energy stored in the inversion is transformed into pulse energy.

8.6.3 Modulators for Q-Switching

The loss modulation of the resonator for q-switch operation can be realized in different ways. Previously, one assumed an ideal quality modulator, which can switch between loss values in an extremely short period of time. Real modulators, however, need a switching time $\tau_s > 0$ for the transition between different states. The ideally given switching conditions are also not always attained: in an open state, residual losses are caused, and in closed state, a residual transparency remains.

The following presents different types of optical switches.

Mechanical Switches
A simple, mechanical modulation procedure is the periodic covering of a resonator mirror by a rotating pinhole diaphragm (Fig. 8.22). The switching times of this

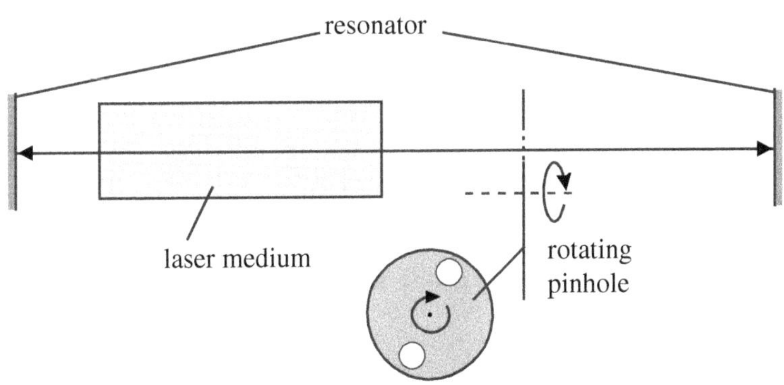

Fig. 8.22 Mechanical q-switching by means of a rotating pinhole diaphragm

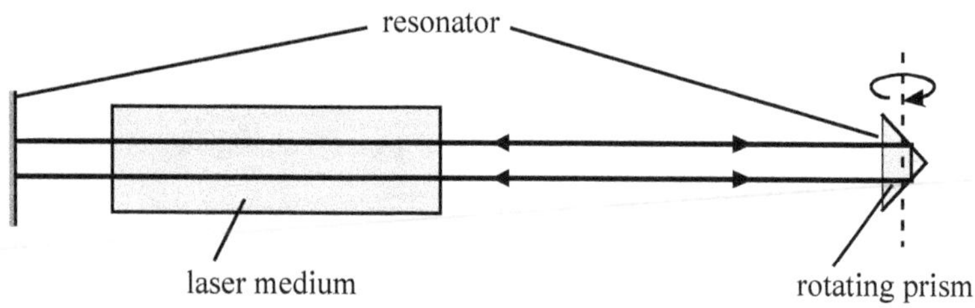

Fig. 8.23 Mechanical q-switching with a rotating prism

procedure are, however, very long and the efficiency is, therefore, small so that it is practically irrelevant for laser applications.

Replacing one of the resonator mirrors by a rotating, totally reflecting prism is a more effective procedure. Only when the prism forms a certain angle to the resonator axis can laser activity begin. This angle setting is only given for a short period of time, $\tau_s \approx 10^{-6}$ s, at high rotational frequencies, and the efficiency is accordingly better (Fig. 8.23).

Electro-optical Switch

The most effective and quickest switches are the electro-optical modulators. For these, an electro-optical element (a Kerr or Pockels cell) is combined with a polarizer. The electro-optical element consists, essentially, of an electro-optical crystal, upon which an electrical voltage can be applied. Under voltage these crystals have the property of being birefringent such that the polarization of the entering light can be influenced by varying the voltage. The electro-optical element has its name from the respective effect that leads to birefringence.

The setup shown in Fig. 8.24 is conceived in such a way that the radiation from the laser medium is initially linearly polarized. If no voltage is applied at the Pockels cell, the polarization remains uninfluenced and the losses are minimal. When an appropriate voltage is applied, the Pockels cell acts like a $\lambda/4$ plate and transforms the linear into a circular polarization. After reflection on the resonator mirror and passing through the Pockels cell again, the polarization is turned by 90° with respect to the initial polarization and is deflected out of the resonator. Thus, the feedback is interrupted. The switching time of electro-optical modulators lies in the range of $\tau_s \approx 10^{-9}$ s, and its switching time can be controlled very well.

Acousto-optical Switches

Sonic waves are produced in an acousto-optical crystal. The periodic density variations produced in the crystal lead to a corresponding periodic change of the refractive index in the acousto-optical crystal. The periodic, varying refraction index now acts like a grid and refracts the incident beam out of the resonator (Fig. 8.25).

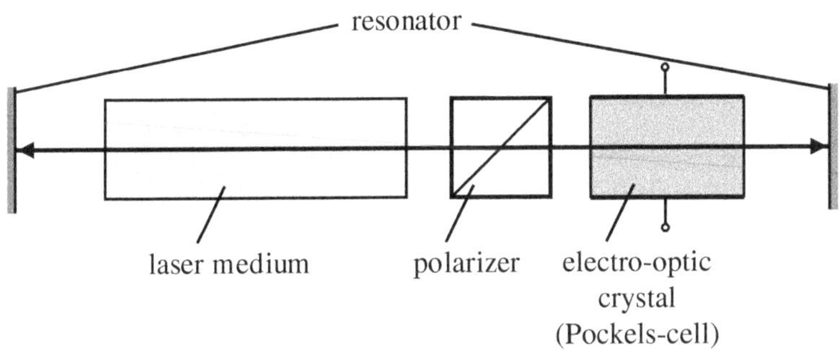

Fig. 8.24 Electro-optical q-switching

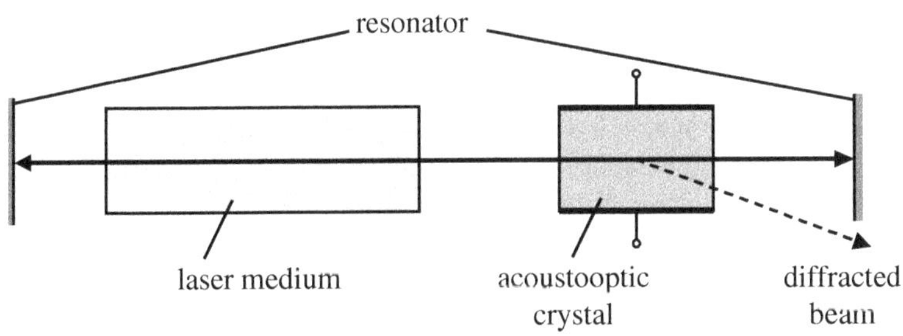

Fig. 8.25 Acousto-optical q-switching

Saturable Absorber

The switches discussed so far are active elements, which mean they are controlled by an external intervention. In addition, there are passive switches, which lead to *self-induced modulation*. The most common and simplest passive switch can be realized by using a *saturable absorber* (Fig. 8.26). The saturable absorber becomes increasingly transparent, the higher the incident intensity is. If the intensity climbs, therefore, over a certain threshold value, the switch is then opened. The absorber has to be selected such that this threshold value is reached precisely when maximum inversion is attained in the laser medium.

Above all, dye solutions or colored glass are used as saturable absorbers. For lasers, certain gases can also be utilized.

In Table 8.1, an overview is given of the typical switching times of the different optical switches.

Table 8.1 Typical switching time of optical switches

Optical switch	Switching time (s)
Rotating pinhole diaphragm	10^{-5}
Rotating prism	10^{-6}
Saturable absorber	10^{-8}
Electro-optical switch	10^{-9}

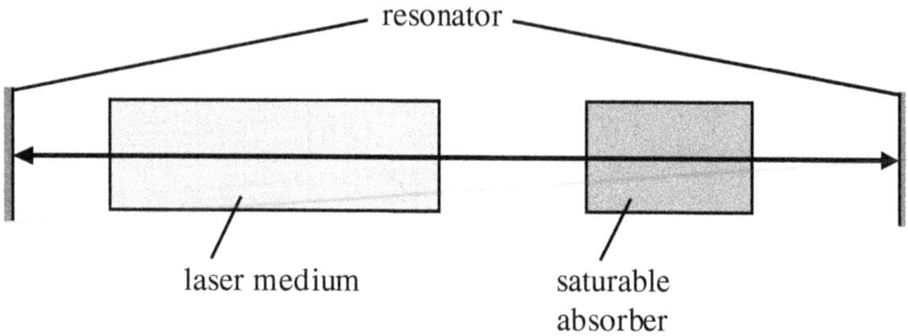

Fig. 8.26 Passive (self-inducing) q-switching with a saturable absorber

8.6.4 Cavity Dumping

Cavity dumping is a procedure related to q-switching which produces short laser pulses. For this, a resonator is needed which can be switched between complete reflection and complete transmission. This can be reached by combining a Pockels cell and a reflecting polarizor (Fig. 8.27).

In the voltage-free case, the (linearly polarized) radiation can pass the polarizer and the Pockels cell and is reflected on the total reflecting front mirror. Due to the missing outcoupling, the resonator losses are very small and in the resonator, a very intensive radiation field is built up. If the Pockels cell, by contrast, is placed under voltage, the polarization direction of the incoming light is rotated by 90°. The light with the rotated polarization direction is reflected on the polarizer and completely outcoupled within one resonator cycle. The cavity dumping procedure is used, therefore, to produce pulses with the length of the resonator cycle time:

$$\tau_{\mathrm{Puls}} = T_{\mathrm{Resonator}} = \frac{2L}{c}. \tag{8.121}$$

τ_{Puls} Pulse duration
$T_{\mathrm{Resonator}}$ Resonator cycle time
L Resonator length
c Speed of light.

The typical pulse duration lies, hence, in the range of 10^{-9} s.

8.6.5 Examples on how to Control the Pulse Form

Through controlling the resonator losses temporally by switching them on and off, both pulse form and pulse length of the q-switch laser can be controlled. A simulation of the laser pulse by gradually diminishing the resonator losses shown

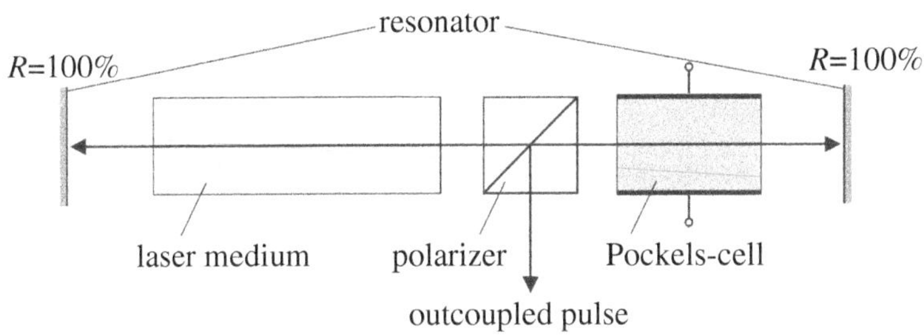

Fig. 8.27 Resonator setup for cavity dumping operation: by switching over the polarization of the Pockels cell, the resonator is opened and the stored radiation energy is outcoupled in a short pulse

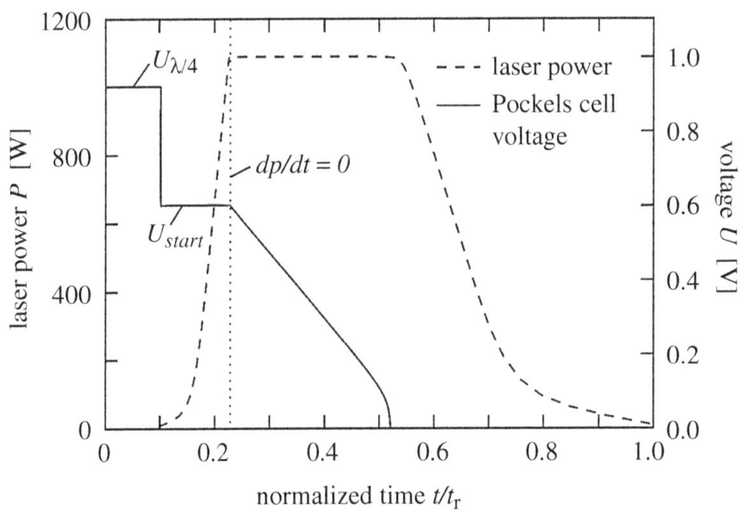

Fig. 8.28 Pulse lengthening by quality modulation. At voltage $U = 0$ the combination of polarizer and Pockels cell is transparent, at $U = U_{start}$ semi-transparent, and at $U = U_{\lambda/4}$ impermeable

in Fig. 8.28 is demonstrated as an example. The increase of the pulse flanks can be controlled by a similar procedure, for example.

A further example is the production of double pulses closely following each other by a two-step modulation of the resonator losses (Fig. 8.29).

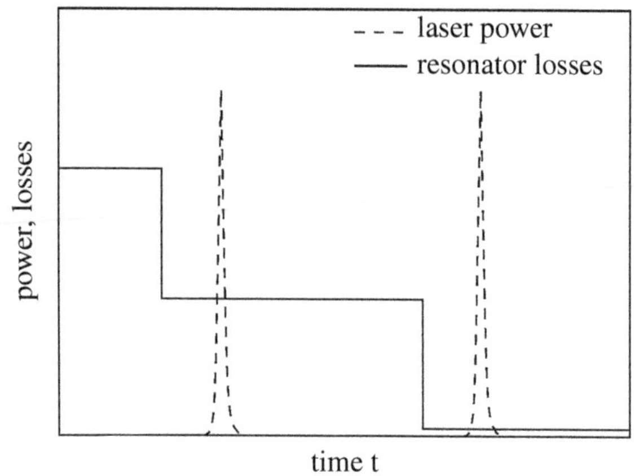

Fig. 8.29 Producing closely consecutive double pulses by two-step q-switching

8.7 Stationary Pulse Generation: Mode Locking

The procedure introduced in the last section to generate short laser pulses are *nonstationary* methods. For this, excitation energy is initially stored in the resonator, which is then briefly re-emitted in the form of radiation energy. The inertia of the system makes the transient and decay time noticeable as an interference and limits the minimum pulse duration. In order to obtain shorter pulses, therefore, *stationary operation* has to be used. This means that the energy stored in the resonator remains constant on a temporal average. Thereby, inertia effects are avoided as far as possible. One method for stationary pulse production is *mode locking*.[9]. This procedure is based on exciting numerous longitudinal resonator modes with a fixed phase relation and can attain pulse lengths of less than 10^{-12} s and pulse power of more than 10^9 W.

8.7.1 Superpositioning of Longitudinal Resonator Modes

The wavelengths of the longitudinal resonator modes (Fig. 8.30) are determined by the condition for standing waves in a resonator:

$$n\lambda_n = 2L, \quad n = 0, 1, 2, \ldots. \tag{8.122}$$

9 .

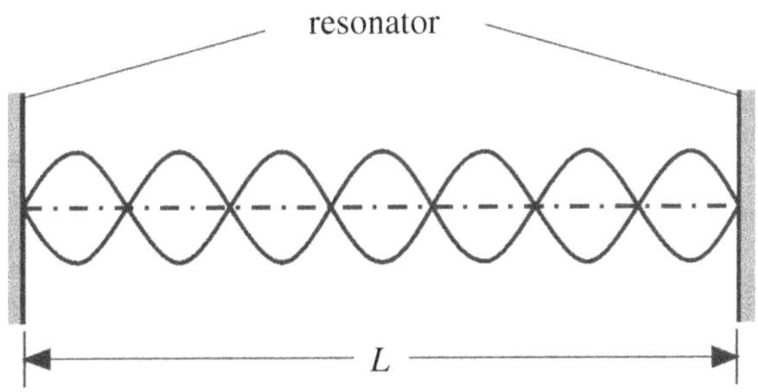

Fig. 8.30 Longitudinal modes: frequency selection by the resonator

λ_n Wave length of the nth longitudinal mode
L Resonator length.

(cf. Sect. 6.1.1).

The angular frequencies ω_n and wave numbers k_n of the eigenmodes are accordingly given

$$\omega_n = n\frac{\pi c}{L}, \quad k_n = n\frac{\pi}{L}, \quad n = 0, 1, 2, \ldots, \tag{8.123}$$

c Speed of light

and the frequency distance of neighboring modes is (Fig. 8.31)

$$\Omega \equiv \omega_{n+1} - \omega_n = \frac{\pi c}{L}. \tag{8.124}$$

Ω Difference frequency of neighboring modes.

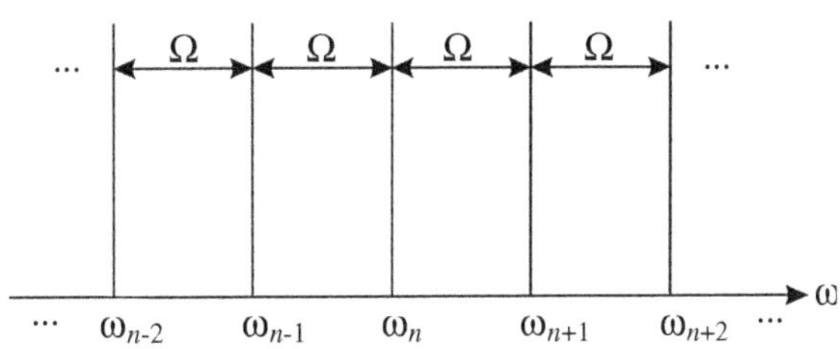

Fig. 8.31 Schematic frequency spectrum of the resonator modes

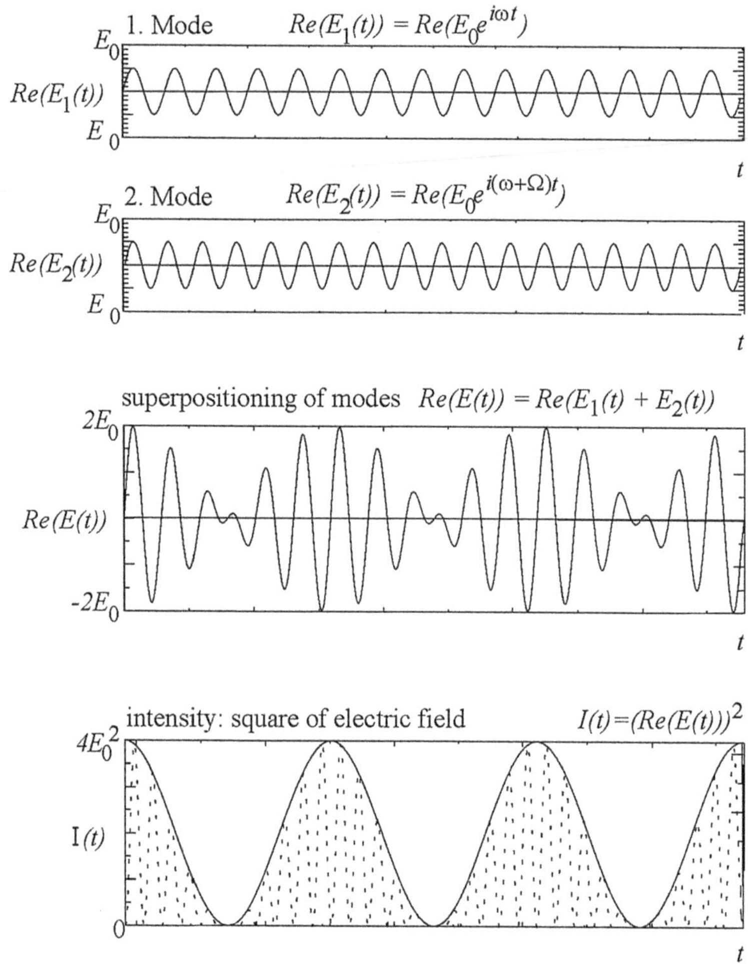

Fig. 8.32 Equiphase superpositioning of two waves with the difference frequency Ω

In the resonator, any number of modes of different frequency can principally exist simultaneously. At first, these modes are independent of each other and, therefore, begin to oscillate with any number of relative phases and amplitudes. The total electrical field in the resonator results from the sum of the field strengths of all modes:

$$E(z,t) = \sum_n E_n(z,t) = \sum_n E_{0,n} e^{ik_n z - i\omega_n t}, \quad E_{0,n} = |E_{0,n}| e^{i\phi_n} \tag{8.125}$$

$E_{0,n}$ Complex field amplitude of the nth mode

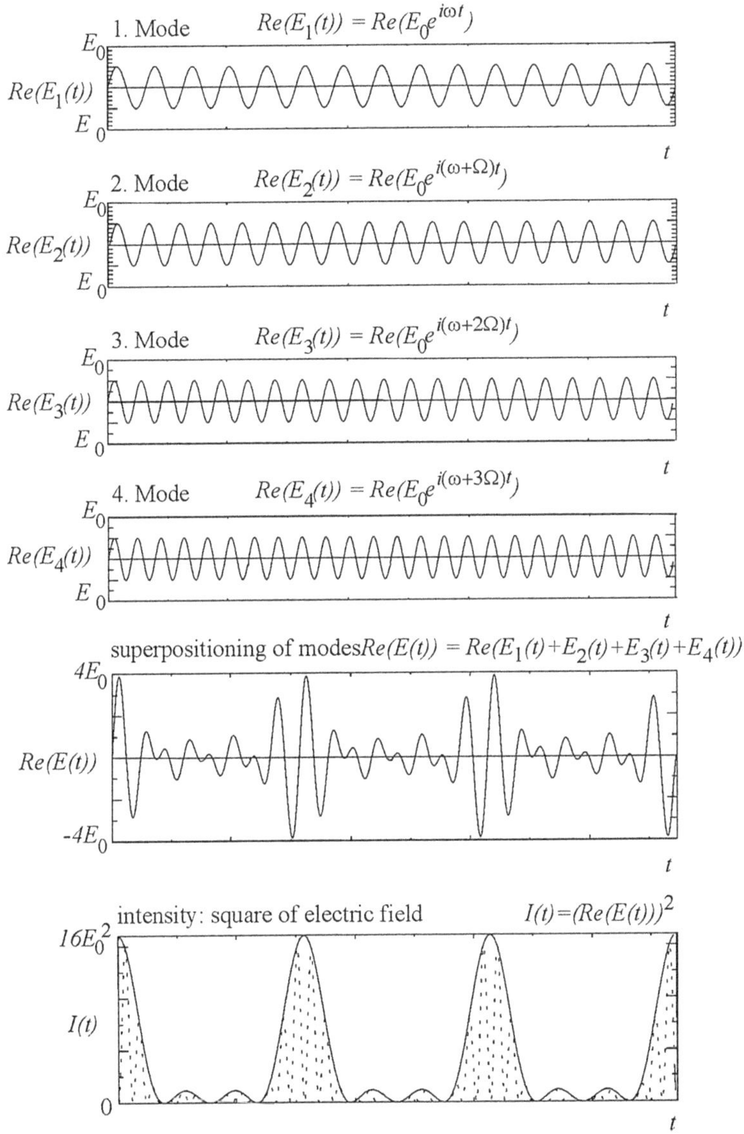

Fig. 8.33 Equiphase superpositioning of four waves with the difference frequency Ω

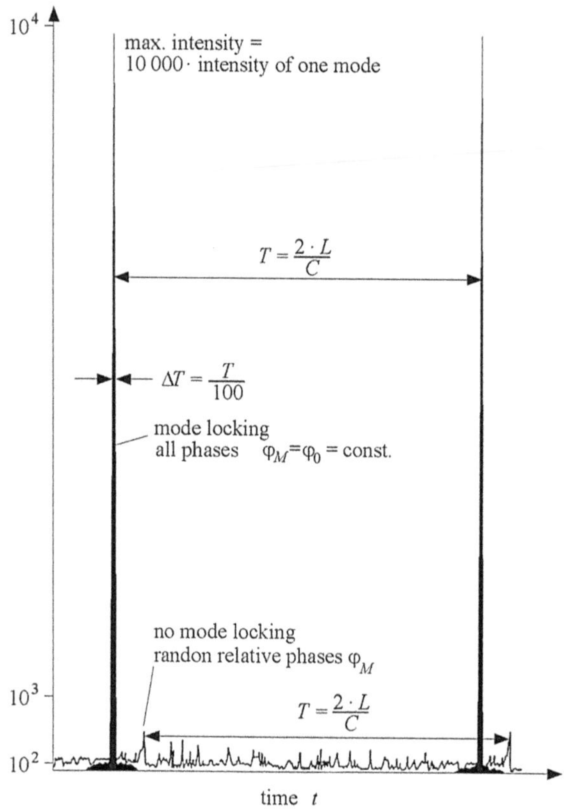

Fig. 8.34 Mode locking of 100 modes

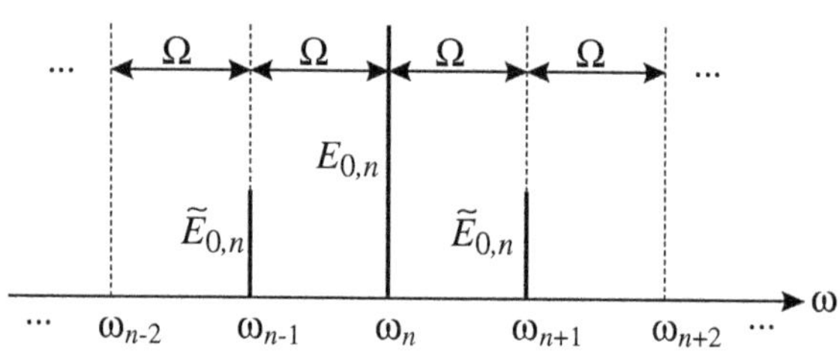

Fig. 8.35 Schematic representation of the sidebands produced by the modulation

φ_n Phase of the nth mode.

Due to the statically distributed phases,

$$\sum_{n\neq m} E_m(z,t)E_n^*(z,t) = 0, \tag{8.126}$$

holds, since the individual summands cancel each other out again on average. The consequence is that the total intensity in the resonator for $N \gg 1$ modes is

$$\begin{aligned} I(z,t) \sim E(z,t)E^*(z,t) &= \sum_{m=1}^{N}\sum_{n=1}^{N} E_m(z,t)E_n^*(z,t) = \sum_{m=1}^{N} E_m(z,t)E_m^*(z,t) \\ &= \sum_{m=1}^{N} \left|E_{0,m}\right|^2. \end{aligned} \tag{8.127}$$

N Number of modes.

Under the simplifying assumption that the amplitudes of the modes are equal, the total intensity becomes

$$\left|E_{0,m}\right| = |E_0| \quad \Rightarrow \quad I(z,t) = \sum_{m=1}^{N} |E_0|^2 = N|E_0|^2 \equiv N \cdot I_0. \tag{8.128}$$

I_0 Intensity of the individual modes.

In the case of statistically distributed phases, the superpositioning of the modes results, thus, in a total intensity essentially constant in time and space, which corresponds to the sum of the intensities of the individual modes.

If the phases of the modes are all equal, then the superpositioning results in a completely different image. Again, the assumption is simplified that all modes exhibit the same amplitude. Then, this time the total intensity results in (Fig. 8.32)

$$I(z,t) \sim |E_0|^2 \sum_{m=1}^{N}\sum_{n-1}^{N} \mathrm{e}^{i(k_m-k_n)z-i(\omega_m-\omega_n)t} = |E_0|^2 \sum_{m=1}^{N}\sum_{n=1}^{N} \exp\left\{i(m-n)\frac{\Omega}{c}(z-ct)\right\}. \tag{8.129}$$

Wherever

$$\frac{\Omega}{c}(z-ct) = 2\pi \cdot j \quad \Leftrightarrow \quad z-ct = 2L \cdot j, \quad j = 0,1,2,\ldots \tag{8.130}$$

is fulfilled, the exponential function in Eq. 8.129 always becomes 1 for all summands. At this point the intensity exhibits a maximum with the value

$$I_{\max} = N^2|E_0|^2 \equiv N^2 I_0. \tag{8.131}$$

The temporal and spatial distance of these intensity maxima can be read from Eq. 8.130:

$$\Delta z = 2L, \quad \Delta t = \frac{2L}{c} \equiv T, \tag{8.132}$$

which means the maxima follow each other at a distance of the resonator cycle time T, and such a maximum is found in the resonator at this point of time (Fig. 8.33).

Due to the fixed phase relation between the modes, regular pulses can be produced with a maximum intensity, which is proportional to the square of the mode number over the intensity of the individual modes. This principle forms the basis of mode locking. As many modes participate, very high peak intensities can be attained (Fig. 8.34).

In order to determine the width of the emerging intensity maxima, the result of the equiphase wave superpositioning is viewed more precisely. For a fixed time, for example, $t = 0$, the superpositioning of the N modes corresponds to the interference of N planar waves exactly. This was already described for the N-beam interference in Sect. 3.7.3. On the screen behind the grid, an interference pattern resulted with the course of the intensity

$$I(z) = I_0 \frac{\sin\left(\frac{1}{2}Nk_z z\right)^2}{\sin\left(\frac{1}{2}k_z z\right)^2}, \quad k_z = k\frac{g}{D}. \tag{8.133}$$

g Grid constants
D Distance between grid and screen
k Wave number of the radiation used.

When transferred to the N mode superpositioning, only k_z has to be replaced by the wave number difference Δk of the modes. In the same way, the temporal course of intensity can be seen for a fixed location, for example $z = 0$:

$$I(t) = I_0 \frac{\sin\left(\frac{N\Omega}{2}t\right)^2}{\sin\left(\frac{\Omega}{2}t\right)^2}, \quad \Omega = \frac{\pi c}{L}. \tag{8.134}$$

The half-intensity width ΔT of the pulse can be determined from Eq. 8.134:

$$I(\Delta T) = \frac{1}{2}I_{\max} \quad \Rightarrow \quad \Delta T = \frac{1}{N}\frac{2L}{c} = \frac{1}{N}T, \tag{8.135}$$

which means the pulse width decreases in proportion to $1/N$ with increasing mode number.

8.7.2 Active and Passive Mode Locking

The last section showed the formation of pulses of high intensity through phase-coupled superpositioning of many modes; the question, however, remains as to how the phase coupling can be reached. A necessary prerequisite for this is a *wide-band amplifying medium* so that as many modes as possible can be amplified in a wide spectral range, which thus are available for mode locking.

All the mechanisms to trigger the mode locking operation are based on the same approach: the resonator losses (or rather, the amplification) are temporally modulated with the difference frequency Ω. The influence this modulation has is easy to recognize when, for example, the reflection coefficient of one of the mirrors is directly changed:

$$r = r_0 + \tilde{r}\cos\Omega t, \quad \tilde{r} < r_0. \tag{8.136}$$

r Amplitude reflection coefficient of the resonator mirror.

Such a modulation of the resonator losses results in an additional time dependence of the field strength of the resonator modes:

$$\begin{aligned} E_n(t) &= \left(E_{0,n} + \tilde{E}_{0,n}\cos\Omega t\right)e^{-i\omega_n t} \\ &= E_{0,n}\mathrm{e}^{-i\omega_n t} + \frac{1}{2}\tilde{E}_{0,n}\left(\mathrm{e}^{-i\Omega t} + \mathrm{e}^{i\Omega t}\right)\mathrm{e}^{-i\omega_n t} \\ &= E_{0,n}\mathrm{e}^{-i\omega_n t} + \frac{1}{2}\tilde{E}_{0,n}\left(\mathrm{e}^{-i(\omega_n+\Omega)t} + \mathrm{e}^{-i(\omega_n-\Omega)t}\right) \\ &= E_{0,n}\mathrm{e}^{-i\omega_n t} + \frac{1}{2}\tilde{E}_{0,n}\left(\mathrm{e}^{-i\omega_{n+1}t} + \mathrm{e}^{-i\omega_{n-1}t}\right). \end{aligned} \tag{8.137}$$

$\tilde{E}_{0,n}$ Modulate part of the amplitudes of the nth mode.

Sidebands are produced over the part modulated with Ω, which overlap exactly with the neighboring modes (Fig. 8.35). Through this, the neighboring modes are excited to forced oscillations at the frequency and amplitude given by the sidebands such that a phase synchronization results of the direct neighboring modes. Since the modulation of the losses concern each oscillating mode, sidebands are produced for each mode, and the coupling extends over the entire mode spectrum.

Basically, the lost modulation occurs similar to that of the q-switch procedure. For the most part, electro-optical or acousto-optical elements are used as active modulators. Then the resonator losses are simply modulated over an alternating voltage of the frequency Ω applied to a Pockels cell. This principle is called *active mode locking*.

Exactly when ultrashort pulses are produced, however, the passive procedure is preferred, which uses saturable absorbers. In this case, the laser pulse circulating in the resonator produces the loss modulation itself: always when the laser pulse passes through the saturable absorber, it is driven into saturation and the absorption losses are reduced. Since $T = 2\pi/\Omega$ is exactly the resonator cycle time, this process causes a periodic modulation with Ω of the resonator losses. This case is called *passive mode locking* or *self-induced mode locking*.

The passive mode locking begins to oscillate in most of the laser systems spontaneously. For this, the saturation limit of the absorber has to be reached at least once so that the loss modulation is triggered. As a rule, the laser process is exposed to enough disturbances and fluctuations that a coincidental intensity maximum reaches the saturation limit in the resonator. If this is not the case and the mode locking process is no longer triggered spontaneously, then such an intensity peak has to be induced by an external disturbance, for example the vibration of a mirror.

GPSR Compliance
The European Union's (EU) General Product Safety Regulation (GPSR) is a set of rules that requires consumer products to be safe and our obligations to ensure this.

If you have any concerns about our products, you can contact us on

ProductSafety@springernature.com

In case Publisher is established outside the EU, the EU authorized representative is:

Springer Nature Customer Service Center GmbH
Europaplatz 3
69115 Heidelberg, Germany

www.ingramcontent.com/pod-product-compliance
Lightning Source LLC
Chambersburg PA
CBHW060746150926
51501CB00003B/103
* 9 7 8 3 6 4 2 0 1 2 3 3 4 *